The Rhetoric of Code: Essays on Technology and Culture

Douglas Thomas

Introduction: The Impact of New Media and Technology

One of the most basic questions that can be broached in any study of technology is what precisely is meant by the term "technology." Perhaps the most intriguing answer to that question was given by Martin Heidegger in his 1955 essay, "The Question Concerning Technology." In that essay, Heidegger writes that "the essence of technology is by no means anything technological."[1] What Heidegger means in that statement is that technology, in and of itself, is not about the latest innovation, the fastest computer chip, or the newest gadget. Instead, Heidegger is making the point that the essence of technology is grounded in the ways in which it mediates human relationships. To that end, it is worth exploring not how only how technology works, but also how technology impacts us in our perceptions of the world and in our interactions with each other.

The Roots of Technology

We can better understand the cultural and social impacts of technology by considering its roots. The word technology comes to us from the Greek, *techne*, a word which, like technology today, has multiple meanings. The two primary meanings correspond roughly with our contemporary notions of *technique*, on one hand, and *technical* on the other. When we think of technique, we often associate that word with a particular stylized, even artistic, manner of doing something. The Greeks classified the arts, such as sculpture, painting, and design as *techne*, insofar as they reflected an individual style or technique of the artist. From that perspective, we can today think of one aspect of technology as the means by which one expresses a creative impulse, the tools which allow that impulse to be made intelligible. In short, technology can be seen as a means of human, artistic expression.

The second idea associated with technology, the *technical*, can be seen less as an individual and creative expression, and more as a set of basic instructions to be followed or a set of tools to perform a certain task. The development of the technical can be traced back to the spread of Athenian notions of democracy throughout the Hellenistic world. In Athens, the art of public speaking was highly valued and young citizens were taught to speak in public from a very early age, spending most of their youth trying to perfect the art. The ability to speak in public was considered to be the core of Athenian democracy. As democracy spread to other Greek cities, citizens of these new democracies found themselves needing to learn quickly how to speak in public. Having little experience, they came to rely on manuals written by the more experienced Athenians, which taught them basic techniques for effective persuasion. These manuals, which were themselves called *techne*, were essentially textbooks, providing basic rules and guidelines for effective speaking skills, allowing people to learn quickly the fundamentals of public speech.

Since that time, most technologies have embodied these two ideas: the artistry of technique and the repetition of the technical. In fact, for a technology to be successful, it needs to be flexible enough to allow for creative, personal expression, while maintaining a set of standards for consistency.

Even with a technology which we consider to be fundamentally artistic, there is always a balance with the technical. While painting, for example, allows for a highly personal form of creative expression (technique), it is also relies on the technical, from naming and classifying art and artists, to the labeling and organizing of art supplies, such as brushes, paints, canvasses, and frames (the technical).

I want to begin by tracing out a very brief (and selective) history of communication technology before introducing some key concepts that will structure the discussion in each chapter. On the technical side, the concepts of the switch and iteration help define the central components of digital media and computing. On the cultural side, notions of transparency[2] and social construction are useful for understanding the social and cultural impact of technology and new media.

A Brief History of Communication Technology

All media technology was at one time new. And with each new, major shift in technology comes a resulting change in the way we think about ourselves, our world and technology. While a complete history of communication technology would require a volume to itself, there are three moments of significant in the evolution of communication technology that are worth reflecting on briefly both because of their impact and because of what they teach us about the nature of technology, innovation and technological change.

Communication technology begins, in a sense, with the alphabet. A system composed of a number of symbols and sounds which could be learned and repeated in a reliable manner forms the basis of writing, translating the spoken word into the written word. The basic principle of all alphabets is iteration: each time a letter is repeated it has (more or less) the same sound. Language evolves from the various combinations of those letters and sounds and made possible a fundamental transition in communication, taking us from speech to writing.

Such a transition, from orality to literacy, transformed knowledge into something that could be separated from people (prior to that singers, town criers, and storytellers *were* the knowledge) and ultimately into something which could be stored, transported and commodified. It also provides an important lesson about the relationship between technology and cultures: knowledge is a key to access. In order to make use of the alphabet, one needed to be able to read and write. As a result, the long standing tradition of orality (such as public readings) would stand in for the new technology (reading) until the skill was widely adopted.

Literacy would lead to a second major innovation, the development of the printing press. The printing press made reproduction of documents possible and with the addition of moveable type by Guttenberg it made it possible to publish virtually anything in a comparably short period of time. The printing press is significant for another reason as well. It illustrates something about the nature of technological innovation and change. The printing press is a synthetic invention, meaning that its innovation is not the result of creating something altogether new, but instead, it is the result of combining other well-established technologies to create something new. In the 15th century, the printing press was created by combining the wine press (one of humankind's oldest inventions) with oil based ink and paper (10th and 11th century inventions). The printing press illustrates what is typically the case with inventions; they fuse previously known technologies into something new.

The third key innovation in the history is the telegraph. The idea of the telegraph was simple, a wire passing a current between two points, but the impact was profound. A signal could be passed between these two points (usually in Morse code) and for the first time near instantaneous communication could occur at great distances. The power of the telegraph was found in its ability to make distance irrelevant. The drawback of the telegraph (and later the telephone) was the way it dealt with bandwidth. Prior to the telegraph, bandwidth was virtually unlimited. Multiple people could read and write using the same alphabet. Multiple copies of books were available so many people could read the same thing simultaneously. With the telegraph, however, bandwidth was restricted. Only one person could use the wire at a time. To deal with the restriction, the telegraph and, later, the telephone, made necessary the invention of the switch, a circuit that could be open or closed and that would allow exclusive access to a wire or line, not allowing others to use it until a call or transmission was completed. So while the telegraph had the power to make distance irrelevant, they also required access and bandwidth, for the first time, to be regulated.

These three epochs in technological innovation provide us with some ideas to understand how new media and technology may be better understood as well.

Computation and Digital Technology

While there is substantial debate over what machine or device constitutes the first computer, there is one example that can serve as a starting point for the discussion of a wide range of issues related to technology and new media. In 1937, George Stibitz, a Bell Laboratories engineer, built a computer out of telephone relays, light bulbs and a cut up tobacco tin, creating the first machine with the three central components that are central to the modern day computer: a processor, a monitor, and a keyboard. Stibitz's crude device was limited in its abilities, it could do nothing more than add single digit numbers together. Stibitz used relays to correspond to numbers and when certain keys were pressed, corresponding relays would switch on, generating the process of numeric addition. Stibitz computers simply turned some switches on and others off, and various combinations of ons and offs were used to represent different numbers.

By using relays to calculate, Stibitz's machine effectively became the first binary computer (Vannevar Bush had created an analog computer a decade earlier). What was revolutionary about it was its digital nature. Fundamental to that nature is the idea of the switch. Throughout its evolution, from telephone relays, to vacuum tubes, to transistors and silicon chips, the nature of the computer has changed very little. Digital computers have grown smaller, faster, and more complex in their design, but their basic function remains unchanged. They are still a series of switches that turn on and off. The number of switches has changed (4.5 million as compared to Stibitz's 2) as has the speed of their operation. The basic design however remains the same.

The second principle that the evolution of digital computers illustrates is the idea of iteration. Computer processors gain speed and processing power not by changing the nature of the switch, but by increasing the number of them. The progression follows a formula called Moore's Law (coined by Gordon Moore, one of the founders of Intel). Moore's Law holds that the number of transistors (switches) on a computer chip will double roughly every eighteen months while the size and cost of the chip remains constant. Computing power is simply a matter of adding more switches, enabling the machine to do more things.

The process of iteration, of repeating the same instruction over and over again with identical results, is also at the heart of modern computing. The switches are coded into a series of states, which represent a particular output, and each time the machine is in that state, the particular result is generated. Modern computing relies on this notion of iteration to function. It is also a fundamental aspect of computer programming as well. Embedded in every computer language is the idea of the loop. A loop is nothing more than a series of iterations, telling the computer to repeat a process a set number of times.

The power of digital computing comes from two very simple ideas: a switch that can be turned on or off and a process which allows for repetition of series of states or calculations. As we add more switches, we are able to make more calculations. As we can make more calculations, we can solve large problems. In short, computers work by taking large problems and breaking them down into a series of very small problems, each of which can be solved in a binary way. The solution can then be recorded by a switch and the solution can be recorded by a series of switch positions or states.

A computer running a program is not much different in how it functions than an election. Consider the example of a ballot measure. Each voter has control of one "switch" and can turn it on to vote yes or leave it off to vote no. The election itself is a process of iteration, with thousands of people turning switches on or off. Nothing more complicated than that happens. At the end of the voting, we assess the state of the machine. If more than 50% of the switches are on, the measure passes. If fewer than 50% of the switches are on, the measure fails. The vote is nothing more than the iteration of a very simple process. Further, we can assign as many states as we wish to the various switches. If 75% of the switches are on, we may call it a landslide. If 100% are on, we may call it unanimous. The number of states is limited only by the number of possible combinations of switches.

The Cultural and Social Meaning of Technology

Technology does not exist in a vacuum. In fact, the ways and degrees to which technology is adopted play a significant role in the process of technological development. While what we do to technology is important, we also need to be mindful of what technology does to us. Arnold Pacey has proposed a tripartite framework for understanding the social and cultural aspects of technology. He divides the general meaning of technology into three domains: the cultural, the organizational and the technical (Pacey, 1983). The cultural meaning of technology, for Pacey, is related to the values that people place on technologies or innovations. The organizational aspect refers to the related infrastructures, ideas, and elements that make possible or sustain a given technology. The technical aspect is the scientific or material conditions of technological production, the actual stuff that makes machines or artifacts function. A very basic example of a car illustrates Pacey's framework. The cultural values we attach to cars, various price ranges, styles, and colors, reflect a set of cultural attitudes about the technology. No one is likely to confuse a red Porsche with a 1973 VW microbus, even though the two technologies serve an identical purpose (transportation). Where they differ dramatically is not in their function, but in their cultural value. The organizational aspect would be all of the supporting structures that must co-exist for cars to function: gas stations, highways, mechanics, traffic laws, even the police. The technical aspect is, among other things, the internal combustion engine. While the engine operates independently of the cultural norms associated with it, its *meaning* is enmeshed in the cultural and organizational aspects which make its production and maintenance possible.

In this sense, we can talk about technology in both the "general" and "restricted" sense. The general meaning of technology concerns all three of these aspects working together, while the restricted meaning of technology is limited only to the technical.

Social Constructivism and Technological Determinism

These two directions also correspond, roughly, to two critical perspectives which have emerged for the discussion of technology itself: social constructivism and technological determinism.

The main tenet of social or cultural constructivism is that humans, as social agents, shape technological change. Technology is viewed as a response to cultural or societal demands. In this view, culture is the arbiter of value and is the force which shapes which technologies will succeed and which will fail and culture is, itself, a complex, multi-faceted network of political, economic, and social forces which interact and behave in unpredictable ways. Technologies which emerge at the "right" cultural moment are likely to succeed and those which appear at the wrong time or in the wrong circumstances are likely to fail.

Technology, according to a constructivist model, is not the product of linear narratives of invention, but rather the result of countless interactive forces which result in the long-term developments of technology. Cultural and social forces are continually shaping and directing the development of technology. The evolution of the printing press, for example, in the 15th century, would be view not as an invention of a single creator (usually wrongly attributed to

Gutenberg), but, instead as a synthesis of a series of prior inventions, such as paper, the wine press, ink, and other forms of block printing. The technologies were in place hundreds of years prior to their reconfiguration in the form of the printing press. For the social constructivist, technology emerges from its surroundings as the result of social and cultural forces, often times synthesizing or recombining earlier technologies to produce new ones.

Technological determinists argue that just the opposite is true, that technology shapes our social and cultural structures. It is a theory which holds that social change is often the product of technological innovation. New technologies shape, alter and define the ways we communicate, behave, and define ourselves. Technology does not evolve, it erupts. Innovation and discovery are the moments of rupture when new technologies emerge and begin to reshape the way we see the world.

We can examine, for example, how each perspective might examine cellular telephone technology. For the social constructivist, the cellular phone is explained by our need and/or desire to communicate with greater regularity and convenience; it is the product of a long series of events and conglomeration of social forces. The cellular phone relies on a vast network of prior technologies: telephones, radio, satellite communication, etc. Social forces were such that the cell phone became necessary and cell phone emerged to fulfill a particular need for human communication. We have cell phones *because* we need to communicate more.

Technological determinism would see the evolution of the cell phone from an entirely different perspective. The technology itself, they would argue, is fundamentally changing the way we communicate. *Because* we have cell phones, we need to communicate more.

The discourse of technological determinism tends to permeate our culture. After a long day of constant telephone annoyances, we may well say "My cell phone is driving me crazy today," neglecting the fact that it is the people who are calling, rather than the technology itself, which is producing the annoyance.

What's New about the New Media?

The idea of new media is based on a fundamental shift in modes of thinking, the shift from analog to digital. That shift has been described in various ways, most famously by Nicholas Negroponte in *Being Digital* as a move from atoms to bits. What Negroponte is referring to is the transformation to an information economy where data which can be encoded in 0s and 1s and therefore transmitted electronically and digitally with become more valuable and more prevalent than its analog equivalent, which remains attached to physical matter. It is the difference between downloading a book on your computer and buying one at the store. The download, free of the physical costs of manufacturing and distribution, exemplifies much of what Negroponte wants to celebrate about the digital revolution, the freedom, access and portability of information. What Negroponte is describing is more a result of the shift from analog to digital than anything about the changing nature of the medium. Moving from analog to digital makes possible the shift from atoms to bits, but it doesn't explain the shift.

We can make a distinction between natural media on one hand and electronic media on the other. A natural medium is one which relies of physical properties of the world to communicate or transmit information. A voice, compressing air, create a wave which hits a listener's ear, vibrates and creates sound. Light waves strike a person's eye and are interpreted by rods and cones into shapes and colors. These are examples of *unmediated* communication. Electronic media transforms those natural waves into electronic waves through a process called *transduction.*

Electronic media also creates two elements missing from natural media: memory and storage. Electronic media have the ability to "freeze" a wave form and replay it. Unlike natural media which are fixed in time and space, electronic media make it possible to record, store and replay images and sounds. Memory fixes an image or sound in a non-volatile or "permanent" state, meaning that it will remain until something acts upon it to change it. While natural media disappear almost instantly (e.g. once something is uttered, it is lost), electronic media provide the ability to store and reproduce sounds and images in a repeatable and faithful form.

The Shift From Analog to Digital

The shift from analog to digital is grounded not in books, music, lights, or computers, but in the structure of electronic waves. Analog waves follow rules based on two parameters: proportion and flow. Increased pressure (or wave stimulation) results in an increase of flow. Analog waves respond proportionally to stimulation, meaning they are like a volume knob on a stereo. When you turn it up a little bit, the volume increases a little, when you turn it up a lot, the volume increases more dramatically. Another device which follows the law of proportion is a car's steering wheel. Turn the wheel slightly to the left to change lanes and much further to the left to make a left hand turn.

Analog's name derives from the idea that analog waves look like the waves they represent. In other words, the electronic waves that represents sound or light, has the same shape and form of the sound wave or light wave that it represents. It is a continuous representation of the form and, accordingly, responds identically to stimulation. You can do the same things to electronic waves that you can to natural media such as light or sound.

Digital waves are completely different. Instead of being electronic representations of waves, digital waves are approximations of those wave forms formed with a series of switches. Each switch is called a bit (short for binary digit) that can have a value of either zero (off) or one (on). Those switches forms a stair step pattern which fills in the space under the wave form to create an approximation. Rather than following rules of proportion and flow, digital waves follow the rule of resolution, a measure of how much variation there can be between various approximations.

If you consider the case of blending two colors together (red and blue) on a spectrum, it is easy to see the ways these two wave forms function differently. When analog waves (light and color in this case) are used, colors fade seamlessly from one shade to the next. There is no way to tell, where the red stops and the blue begins, you are simply increasing the flow of red, which results in a greater proportion. Think of it as slowly and continually pouring red paint into a can of blue paint as you stir the two together.

As a result, analog waves, for the most part, can only be accessed in a linear fashion. The have beginnings, middles and ends, and in order to access them in an intelligible way, you must start at the beginning, proceed through the middle, and finish at the end. Even technologies such as VCRs follow this rule. When you skip over the commercials in your favorite recorded television program you use “fast forward,” a process which accesses the content linearly, but at a much faster rate, appearing to skip over it.

In contrast, if we return to our paint mixing example, such a digital transition is marked by the colors passing through a successive series of states. Each state is discrete and is able to be turned on or off. The measure of resolution is a description of how much variation I can have between two adjacent states.

If my resolution is 4 bits, that means I have four switches which can represent my transition. Accordingly, I would need one state which is all blue, one which is all red as well as series of fourteen other states (4 switches yield 16 total possible combinations or states). Instead of a smooth flow from the first state to the last, I would have fourteen separate and discrete blocks which I would move through to represent the transition.

At very low levels, such approximations are crude, but the power of digital media is grounded in the process of iteration. Adding more switches increases the resolution dramatically. Resolution is governed exponentially, meaning that each switch you add increases the number of possible states by raising the exponent, rather than the base. One switch is 2^1, meaning it can have two states, on and off. 2^1=2. Adding a second switch raises that exponentially. Two switches is 2^2, giving us four states: off/off, on/off, off/on, on/on. A third switch, has 2^3 or eight possibilities, and so on.

This formula provides the measure for a resolution: 4 bit resolution provides for 16 possible states, 8 bit resolution provides for 256 states, 16 bit resolution provides for 65536 states and 32 bit resolution provides 42,949,667,296 states. If you think of those numbers in terms of available colors on a computer screen, a 4 bit picture has a palate of only 16 colors in order to render an image, which a 32 bit picture can select from a palate with more than 250 million times that many options.

Resolution relies on the power of iteration. Adding more switches provides the computer with an exponentially increasing array of choices or states to more through, allowing greater and greater approximations of wave forms.

The waveforms that are generated, however, are not continuous. Because they are marked by a series of discrete states, the beginning, middle and end or each waveform is arbitrary. As a result, digital media are able to produce non-linear artifacts which provide several advantages over analog storage. First and foremost, digital waves, since they can be accessed linearly, provide the same storage benefits that analog waves do, but because they can be accessed non-linearly, they provide additional means of access. Digital media can be read through what is called *random access*, meaning that any random point is just as accessible as any other (not that they are accessed randomly). That means that at high resolutions, you can access any particular state, rather than having to move from beginning to end to find information.

Predicting the Future

There can be little argument that technology and new media impact our lives in dramatic and significant ways. They change our behaviors, our patterns of living, the way we communication, even who we communicate with. Technology and new media also serve as a means to project our hopes and fears into the future as well as reflect on the present. In that sense, technology plays host to both our greatest dreams and our darkest fears.

Historically, nearly every modern technology has been met with a set of conflicting visions. On one hand, technology promises to change the world for the better, to usher in a new age of convenience, hope, and salvation. These narratives are utopian in nature. On the other hand, each of these new modern marvels also spawn a host of more dire predictions, predictions of a world where machines dominate humans, privacy is eliminated, and social and cultural mores are radically altered for the worse. The narratives describe a dystopia, where technology runs amok.

Both of these narratives are born out of a technologically deterministic viewpoint, suggesting that technology is autonomous, self-reproducing, and to some degree inevitable. Where they differ is in the impact and values that each ascribes to the growth of technology.

Utopia is a term coined by the 16th century philosopher Sir Thomas More in which he outlined the conditions for the ideal political state. Literally translated from the Greek, Utopia means "no place." More's traveller, Raphael Hythlodaye, gets lost en route to America and comes across an island named Utopia where he finds a political paradise in operation. The book reflects both on the dream of a More's idea of what government should be, as well as providing a critique of the monarchy of his time. The idea behind utopia is based on the ability to clearly envision what seems to be impossible and to commit oneself to achieving these impossible goals. Ideals such as world peace, the end of world hunger, and elimination of poverty are all utopian in the sense that it is relatively easy to imagine such an outcome, but difficult to articulate a plan of action to achieve it.

Utopian technological narratives provide an understanding of the means by which we can achieve what appear to us to be impossible dreams. Accordingly, they can be understood through the lens of *instrumentality*, which is to say, they view technology as an instrument for human progress and betterment. One of the fundamental premises of such instrumental visions is that order is, essentially, a positive force. Technology, therefore, contributes to human advancement by making our lives more ordered and therefore more productive and more clearly defined.

Such ordering assumes that technology is a product of human design and manufacture and accordingly that technology exists in the service of humankind. These narratives also assign a moral or ethical neutrality to the technology itself, meaning technology can be used for good or ill, but the moral and ethical dimensions of such use remain with the people who use it, not the technology itself. Finally, such narratives assume a certain degree of control over technology, namely, that because it is our invention and its value it largely determined by how we use it, technology is something that is able to be mastered by those who create and use it.

Utopian narratives often assign authorship or invention to technology as well, constructing stories of moments of innovation. Utopian thought relies to some degree on visions of technology which provide a fictional justification as well as a scientific strategy for achieving the goals it puts forward. In doing so, utopias are often guided by the vision or plan of a single individual (although one could argue this is rarely actually the case).

For example, Bill Gates is often credited with revolutionizing the personal computer market with his vision of "a computer on every desktop." While Gates certainly did have a key role in making that happen, he was merely a part of that revolution, which would not have been possible without a host of other people, companies, and ideas that helped Microsoft gain market dominance in the 1980s. From a utopian perspective, however, it illustrates the structure of such a narrative. Gate's operating system was seen as the instrument of transformation and he is given credit for authorship and invention (neither of which is accurate) for the transformation that resulted. In the utopian version, Gate's vision, "a computer on every desktop," was the commitment which fused the fictional ideal with the technological invention to bring about change.

The dark side of technology also has its own set of narratives, mainly born from resistance to change and the looming specter of machines replacing humans. Although dystopian narratives have likely accompanied most innovations, the roots of contemporary dystopian narratives took hold during the industrial revolution. With the birth of the factory and mass assembly, most of the West was transformed from a system of artisans and apprentices to one which relied on machinery and unskilled labor. Machines were now capable or working more quickly and more efficiently than craftspeople, resulting in a wider availability of good at lower prices. Labor, now unskilled, also became available at lower wages, resulting in mass production of consumer goods.

As the industrial revolution took hold in Britain, there emerged a group of workers dissatisfied who resisted the onslaught of machinery and factories. In order to combat industrialization, the created a fictional character named Ned Ludd, who was to lead a revolution against the machines. Followers of the movement became known as "Luddites." Luddites would find ways to break machinery and subvert the manufacturing process. When a machine would break, usually at the hand of a worker, they would complain to management that Ned Ludd had caused the damage.

Ned Ludd and other more contemporary incarnations of luddite behavior focus on basic fears of technology encroaching into the workplace and the home. Often these fears are accompanied by anxiety about being replaced by machines or of humans having their value determined in contrast the machinery which can function more efficiently then they can.

Today, as technology become more ubiquitous, concerns have enlarged to encompass nearly every aspect of human activity from the largest (concerns about government and privacy) to the more basic (fears of technology causing depression or anxiety in individuals).

Visions of Technology

By most accounts, US science fiction began with Hugo Gernsbach's *Amazing Stories* in 1926. *Amazing Stories* published stories that embedded notions of science into fictional tales and in doing so invented a new literary genre: science fiction (its original subtitle was "The magazine of scientifiction") (Nichols, 26). In the past 80 years, science fiction has undergone a number of transformations and the genre has been variously expanded and contracted to include works which bear little resemblance to the early works of "scientifiction" that Gernsbach published, but throughout a set of constant themes have remained central to most works of science fiction.

Science fiction has emerged as both a barometer of the public imagination and as a place we look to give shape to our greatest hopes and our darkest fears. It entertains us (roughly half of the top 10 grossing films of all time are science fiction, depending on how you count). But it also warns us of worlds that threaten to become our own, if we are not mindful of the darker side of human nature and technology (Orwell's *1984*, Bradbury's *Fahrenheit 451*, or Huxley's *Brave New World*, for example). Finally, science fiction shows us the promise of the future and gives rise of fantastic dreams (Asimov's *I, Robot* stories and Jules Verne's *A Trip to the Moon*).

What makes science fiction different from other genres of literature is that science fiction stories are grounded in science. There is great debate about the nature, scope and meaning of the genre, especially in its present form, but three of the most important influences on the shaping of science fiction were, undoubtedly, Jules Verne, H.G. Wells, and Mary Shelley. Each exemplifies a key facet of the genre of science fiction. Verne's work speaks to the ways in which science can be used in exploration and discovery. Wells used science fiction to issue warnings and provide cautionary tales, and Shelley's classic story *Frankenstein* written in 1818, remains, even today, the most pervasive and most fundamental model of scientific critique.

Notes

[1.] Martin Heidegger, "The Question Concerning Technology" in *Basic Writings*.

[2] I borrow this term from Sherry Turkle's *Life on the Screen*.

Chapter One: Innovation, Piracy, and the Ethos of New Media

With the development of broadband Internet connections and increasingly sophisticated means of compression, new media is radically reshaping our communication environment. The emerging environment, however, is the product of two distinct sets of technology. On one hand, 'old' media, from the age of print through broadcast and film has a long-standing corporate tradition, and are clearly regulated both in terms of issues of content and distribution. Issues of ownership, marketing, distribution are well understood and have decades of legal precedent demarcating boundaries. On the other hand, 'new' media, which focuses primarily on transformation from analog to digital is too recent to have a well-established tradition of regulation and is too different from its predecessors to be easily regulated by older laws.[1]

Regulation, old media proponents argue, is necessary to protect copyright, ownership, and intellectual property rights. New media proponents see regulation as a barrier to innovation, dismissing more traditional notions of intellectual property as an outdated way of thinking about information. The tension between old media claims of piracy and new media claims of innovation reveal two different perspectives on the same phenomenon, primarily sparked by the Internet and World Wide Web's ability to distribute information instantaneously on a global, public scale.[2] What new media champions see as innovation in communication, old media proponents are likely to see as a threat the stable system of information management and distribution which has long regulated music, film, and broadcast economies. The problems multiply as more forms of old media are transformed into digital content and made available on the net.

This chapter is an attempt to understand both piracy and innovation by addressing the historical trajectory that new media has taken, particularly the development of computer software, in order to define an ethos of new media. In short, the history of new media production, from the earliest forms of software for the personal computer, to video games, sound, and full motion video files, can be understood more fully by examining the context of development and innovation in which these forms of communication emerged. That history helps explain not only how new media functions, but also how mainstream culture has come to think about new media and how we have come to resolve the long-standing tensions between piracy and innovation.

Who Owns Information?

Of all the questions that new media has raised, perhaps the most hotly debated is the question of intellectual property, the question 'Who owns information?' Nowhere has that debate been more fiercely fought than around notions of 'piracy' on the Internet. The question of WWW piracy is often reduced to a matter of financial loss. Microsoft, in fact, goes as far as to argue that the impact of piracy, in addition to causing 'higher prices', 'reduced levels of support' and 'delays in the funding and development of new products' also harms 'all software publishers' as well as 'the local and national economies' resulting in 'lost tax revenue and decreased employment'.[3]

Those who write, market or sell software, music, or film, for example, see piracy as cutting into their bottom line. At the most basic level the argument is irrefutable. Those who pirate software or entertainment media are not paying for it. As a result, it is tempting to think of piracy as theft and that provides an adequate explanation both for the casual 'pirate' who would prefer to not pay if he or she doesn't have to, as well as large scale piracy operations where thousands of video cassettes of the latest film or CD-ROMs for the latest software titles are made and sold on Manhattan street corners.

There are important elements of the new media landscape that complicate such a simple formulation. New media has shifted the idea of intellectual property onto new ground. Instead of considering information from the vantage point of content creation (the essence of old media), new media invites us to think in terms of two different concepts: reproducibility and distribution. In terms of old media, our laws of copyright are straightforward in assigning ownership of property, even intellectual property, to those who create or produce information. In terms of new media, however, the landscape appears to be shifting. Because of the digital nature of new media and the availability of extensive networks, those who purchase or otherwise obtain new media, almost from the moment of purchase, are poised to become distributors as well as consumers.[4] That transformation has altered not only the way in which we purchase media content, but also the way we think of it as reproducible.

Whereas old media was reproducible in terms of content, the act of reproducing it often resulted in loss of quality (e.g., taping an album or duplicating a video tape). By way of contrast, because information in new media is reduced to 'code', it is capable of nearly instantaneous, unlimited duplication with no loss in quality or content. Old and new media, then, follow two different logics of reproduction and as a result, two different ethics for distribution.

Logics of Reproduction: From Art to Code[5]

Understanding what is at stake in a discussion of the impact of technology on intellectual property, ownership and piracy necessitates rethinking our basic notions of what it means to reproduce digital media. The traditional notions of reproduction, based on the idea of a copy of an original, begin to break down in the face of digital transformation. This transition is marked by the logics of two different systems of reproduction. The first, I refer to as an 'artistic' logic of reproduction. The second, marked by the introduction of digital technology, I refer to as a logic of 'code'.

The idea of artistic reproduction is drawn from the notion of representation, literally the re-presentation of an image, object or idea and is animated by the concept of difference. It is a logic sustained through the idea of the copy. The notion of a copy in Western thought, since the time of Plato, has been based on a differential relationship between an original and a reproduction of that original.[6] That sense of duplication is in fact defined by the degree of difference interjected into the reproduction and the degree to which the copy fails to correspond to the likeness of the original (or alternatively, the degree of resemblance to the original and minimization of difference). Those notions of similarity and difference become the basis for rendering judgment about all forms of reproduction. These degrees of similarity, and particularly difference, are responsible as well for what Gilles Deleuze has called the 'process of individuation', the means by which reproduction already presupposes a form of difference which makes repetition (the production of an identical copy) impossible.[7]

It is in this sense that Walter Benjamin discusses the reproduction of the work of art in what he has deemed the 'age of mechanical reproduction'.[8] The notion of difference, for Benjamin, stems from the possibility of spatial and temporal dislocation. 'Even the most perfect reproduction of a work of art', Benjamin argues, 'is lacking in one element: its presence in time and space, its unique existence at the place where it happens to be'.[9] It is that grounding, both in a physical place and a fabric of tradition which gives art its value. Accordingly, the value of art, for Benjamin rests with what is un-reproducible (what he defines as an 'aura'), qualities which are unique and inseparable from the object in its status as an original. Mechanical reproduction, Benjamin contends, destroys that aura: 'By making many reproductions, it substitutes a plurality of copies for a unique existence'.[10]

A second logic, that of code, emerges as a function of digitization. The digital is literally a coding of information into 1s and 0s with the goal of making perfect reproduction possible without difference or loss. As a moment of perfect reproduction it removes the relevance of difference in the determination of judgment. In short, the distinction between the copy and the original is no longer able to be judged and as a moment of perfect reproduction, the copy and the original become indistinguishable. Unlike artistic representation, the logic of code does not preference the original over the copy. Instead, the original is seen as already multiple, designed for reproduction and distribution. Once complete, there is no material way to distinguish the copy from the original.

This second logic of reproduction, then, confounds what we know about systems of reproduction. Once copies become indistinguishable from originals, reproduction loses a sense of authenticity. In order to maintain a standard for judgment, something else must substitute for the ability to distinguish between copies and originals, should that distinction be seen as an important one to maintain.

Digital reproduction is not animated by a distinction between a copy and its original, but, instead, by a sense of authority. The notion of the copy has been transformed from an object to an activity. In fact, the very definition of the copy in the digital age makes no mention of the relationship of the copy to an original. According to Microsoft's 'End User Licence Agreement' (EULA):

> You make a 'copy' of a software program whenever you: (1) load the software into your computer's temporary memory by running the program from a floppy disk, hard disk, CD-ROM, or other storage media; (2) copy the software onto other media such as a floppy disk or your computer's hard disk; or (3) run the program on your computer from a network server on which the software is resident or stored.[11]

In the digital age, the copy is defined by the act of loading, copying, or running a program. Nowhere does it specify the relationship of a copy to an original. Instead, the distinction, and therefore the basis for judgment, is grounded in <u>authority</u>, literally the conditions under which reproduction is allowed. The definition of the copy is still differential, but now the relationship is expressed as a ratio of who may and who may not engage in the act of copying. The copy, for Microsoft, is therefore defined as matter of the <u>right</u> to reproduce software, rather than in terms of the reproduction itself. Accordingly, Microsoft's definition of piracy is grounded in activity, rather than product:

> Software piracy is the unauthorized copying, reproduction, use, or manufacture of software products. On average, for every authorized copy of computer software in use, at least one unauthorized or 'pirated' copy is

> made. In some countries, up to 99 unauthorized copies are made for every authorized copy in use.[12]

The difference between piracy and software production and distribution rests not with the activity of copying. The actions of the pirate and Microsoft itself are identical. Both transfer bits to storage media in precisely the same way and each produces a product identical to the other. The distinction is no longer a matter of quality of reproduction.

In essence, the world of the image and text, the world of representation, has been transformed in the digital age into a world of code, of 1s and 0s, easily repeated and distributed through ever-increasing communication networks. Because the relationship between the copy and original no longer serves as the basis for judgment, content no longer serves as a valuable basis for judgment or evaluation. There simply is no distinction between the copy and the original and no way to distinguish between them in a system of digital repetition.

Judgment is, therefore, displaced from the object reproduced to the activity of reproduction. That activity is defined as the movement of information (bits) from one place to another, whether it is from a disk to the computer's memory or from one computer to another. In short, reproduction, as a function of movement, has become synonymous with distribution. As a result, piracy and ownership in the digital age, from software to emerging forms of new media, are more about the right to distribute than the right to reproduce information. In what follows, I trace out a series of cultural moments which have help to shape and define the ways in which digital information has come to be regarded differently from their more traditional analog counterparts and how what is seen as innovation from one perspective is seen as piracy from another.

A Brief History of Digital Distribution

I want to emphasize two distinct moments in the history of digital distribution that have shaped an ethos of the digital community: the development of peer-to-peer file sharing and the growth of underground pirate communities (warez boards) which emerged in the late 1980s. Each moment delineates not only a set of circumstances under which a new form of digital distribution became possible, but also an ethic that accompanied that form of distribution. That ethic emerges from the idea that technology subcultures, while often times operating outside the law or in direct confrontation with it, are generally heavily self-policed. Because the law tends to do a poor job of regulating new media effectively (or wisely), those who work in the environment tend to engage heavily in self-regulation. Accordingly, the history of new media is also the history of an ethic, which has developed in response to new forms of communication media. Understanding that history is crucial to understanding how it is that old media faces new problems in the new media landscape. Each of these moments documents a critical point at which the question of distribution has conflicted with a set of broader corporate interests and in doing so has called into question the basic logics of capital, which underwrite notions of control and, ultimately, ownership of information. As more classically corporate forms of media (music and film in particular) enter the digital landscape, which is to say as they are transformed from 'art' to 'code', they enter a new domain where issues of content distribution have a radically different history.

Perhaps the most significant 'ethos' to develop in relation to systems of digital distribution grew out of peer-to-peer file sharing systems which emerged in the late 1990s. Software such as Napster (and its increasingly decentralized cousins Kazaa, Morpheus, Bearshare, and Gnutella) extended the idea of information exchange to a new level. With programs such as Napster, commercial music (and now film and video) where able to be freely exchanged online. Virtually any song, album, or performance desired could be found in digital form and downloaded to a user's machine free of charge. The significance of Napster rests less with the technological advances of peer-to-peer file sharing then it does with the ethos that it spawned among computer users.

As the world of free music opened up, users began to co-opt the earlier hacker ethic, which had given rise to most of the technology that was now begin used to allow free file exchanges. This ethos grew out of one of the earliest mantras of the computer community, the idea that 'All information should be free'.[13] But when this generation of computer programmers and enthusiasts spoke of freedom in the 1960s and 1970s, they were not thinking of it strictly in a financial sense. They were speaking more about the free flow of information. Anything which made access to information more difficult or which blocked the transmission of information was considered undesirable. What was at stake for these programmers was not software that was 'free' financially (though most of it was cost-free), but the freedom to explore, alter and improve on the software.

The Napster ethos translated that ethic in the most basic, and, perhaps, most convenient terms. But it is important to note that what music company executives and some artists saw as stealing, many users considered to be free. As a result, the Napster ethos has three inter-related elements: corporate resistance, freedom to redistribute, and entitlement to digital content.

First, on what is perceived as an increasingly corporate Internet, users see trading of MP3s and digital music as an act of resistance against precisely the corporate agents who seek to control and regulate the Internet. In that sense, at the most basic level file trading is seen as an act that embodies the spirit of the Internet, keeping information free and open in the face of corporate control. Second, a large percentage of users who trade on Napster-like sites feel that they have the right to re-distribute music they have purchased, provided they don't profit financially from the act. The analogy is often made to taping albums for friends, trading tapes, or recording songs from the radio. The only difference, they argue, is the technology being used. Third, and perhaps most important, many users feel an entitlement to digital content based both on a presumption that anything appearing in digital form in public is (and should be) free, as well as the sense that the intermediate technology has already been purchased. In short, after paying $2,000 for a computer and between $20.00 to $40.00 per month for Internet access, many users see the content as already paid for in their monthly network access fees. Moreover, after paying an additional $13.99 for a CD, they feel they have the right to duplicate and share it as they see fit.

These three elements, corporate resistance, freedom to redistribute, and entitlement to digital content, have created a sense of how digital information should flow and have marked the site of redistribution as the place in which users can battle corporate control of the Internet. As the means of distribution becomes increasingly decentralized, the corporate focus will also need to shift toward a system of distribution that makes digital reproduction increasingly difficult as well.

The first shots in that battle have already been fired with the passage of the Digital Millennium Copyright Act (1998), which states that 'No person shall circumvent a technological measure that effectively controls access to a work protected under this title'. In effect, the DMCA makes it illegal provide tools, information, or technological devices, which allow any form of copy protection to be broken or which allows any form of unauthorized reproduction. One of the first tests of the DMCA has come as a result of a link posted to the 2600 Website (the Website for the hacker magazine of the same name). In 1998, 2600 posted a link to DeCCS, a software algorithm that was capable of decoding DVD recordings, allowing them to be reproduced and distributed.

The battle is one that pits the traditional hacker ethic of exploration and free sharing of knowledge and ideas against corporate interests to protect their products. The response from industry (and government) illustrates their misunderstanding of the problem of distribution. They believe if they can make their music or film secure (by prohibiting the dissemination of information which would make it insecure), then they will not have to address concerns about redistribution, duplication and piracy. What the industry is failing to account for is the fundamental transformation that has occurred in the medium of dissemination. That is to say, that while hackers have adopted the logic of code, industry is still fighting the battle from the perspective of art.

The problem that industry faces stems from its reliance on traditional models of distribution, where one needed to purchase an album, CD, VHS cassette or DVD in order to consume their product. Today's model of digital distribution challenges the primacy of the physical medium and in doing so sets a new dynamic in motion. The difference between these two models of distribution makes note of the distinction that Nicholas Negroponte uses to introduce his work Being Digital, the distinction between 'bits' and 'atoms'.[14] Bits parallel the Napster, peer to peer, Open Source model, the digital encoding of information into 1s and 0s which flow effortlessly on the information superhighway (and for the most part Open Source software is available online, rather than in a physical medium). Atoms, the material stuff of the world, follow a different set of rules. Unlike bits, which are nothing more than information, atoms have physical presence and it is that idea of physical presence that facilitates notions of ownership as the concrete possession of an item or, by extension, an idea. Entertainment content and software, which had traditionally been sold as atoms (media such as floppy disks, CD-ROMs or DVDs), is now becoming more widely available as bits (through direct download on the Internet), particularly as bandwidth increases.

This moment, then, is defined by conflicting modes of distribution, one that facilitates freedom and openness (and embodies the Napster ethos), the other which attempts to maintain rights based on exclusive possession and ownership (the corporate model). As the computer began to take on a heightened role as a communication medium in the 1990s those notions of distribution would, once again, clash. As a result, out of the computer underground would emerge a new group of hackers with an entirely new ethic.

In the mid 1990s, after the formation of the Internet, but prior to its widespread public use, small groups of self-described pirates began utilizing a network of small, private BBS (Bulletin Board Systems) to trade and distribute software. These software traders (usually high school aged boys) prided themselves on cracking the copy protection on software the day it was released (or even sooner if they are able to get access to pre-release copies) and would distribute that software via an underground community on self-described 'Warez' boards. These warez traders were (and continue to be) relatively few in number and tended to provide access to software to other high school aged kids, providing access to games or to applications that they would not be likely to afford or purchase independently.

The focus of these groups, with names like RiSC (RiSE iN SUPERiOR COURiERiNG) and PWA (Pirates with Attitudes), was not on their skill as programmers or even copy protection crackers, but in their ability to distribute the software within hours of its release. Advertised as 'zero day' or even '0-12 Hour Warez' the programs would be transmitted via modem to BBS all over the country, which allowed other users who had gained membership the ability to access programs. BBS were chosen by the groups and given varying degrees of affiliation, ranging from 'World and US Headquarters' to 'Member Sites'. In most cases, the BBS systems employed a quota system, measuring the user's contributions as well as their downloads. Those not contributing their fair share were branded as 'leeches' and were either banned from the board or would have their account name published on the board in an effort to publicly shame them.

The BBS as a distribution medium prefigured much of what would follow with the Internet, initially with FTP sites and later with trading on Internet Relay Chat (IRC) and, eventually, the WWW. In 1992, Bruce Sterling described the BBS as 'a new medium . . .even a large number of new media' with unique characteristics:

> Boards are cheap, yet they can have a national, even global reach. Boards can be connected from anywhere in the global telephone network, at no cost to the person running the board. . . .Boards do not involve an editorial elite addressing a mass audience….And because boards are cheap and ubiquitous, regulations and licensing requirements likely would be practically unenforceable.[15]

Within this context, it is easy to see how unauthorized software could be distributed quickly, at little or no cost, with great incentive to reach a national or even international audience as a matter of reputation. Moving from BBS to the WWW was only a shift in the degree rather than the type of distribution.

The emphasis with pirate BBS or pirate groups was on the speed with which they were able to make software available, not on the cost-free nature of the software itself. In fact, many of the pirate groups would specifically disavow any financial motivation. Both RiSC and PWA would add information files (marked as files with an .nfo extension) to the software they distributed, often times providing information about the software and its installation, but more often as a form of advertising for the group. The .nfo files would tell the end user who was responsible for the cracking and the distribution of the software.

Perhaps the most interesting element and notable element of the warez ethic has to do with their constant and specific disavowal of financial motivation. Cracking copy protection for profit is anathema for warez traders. In 1996, PWA's .nfo file would include the following disclaimer: 'Please note that PWA is NOT accepting pay sites of any nature. We're in this for fun and entertainment, not to try to make ourselves rich'.[16] RiSC would add a similar statement in the .nfo file:

> RiSC is the longest lasting courier grp by far, and continues to bring honor and respect to the courier scene as only RiSC can. RiSC does not take donations of any sort for our services. We work on merit alone, the way the scene should be run.[17]

The soldiers of any pirate group were not the programmers, but usually help the title 'courier' and it became their responsibility to spread the cracked software as quickly and as widely as possible. The ethic that emerged in the warez underground is similar to the ethic that has always driven computer enthusiasts: to make quality software immediately and completely accessible.

That ethic has extended to other forms of new media as well, with pirate and warez groups setting up anonymous WWW and FTP sites (often using free server space) to distribute digital music and video. Because these accounts are routinely deleted by service providers, the knowledge of distribution networks becomes essential for warez traders. Pirate copies of software, music or video are short-lived (often lasting less than 24 hours on a server), so warez trading continues to remain a question of access, providing those who are 'in the know' with virtually unlimited access to digital music, film and software.

Everything Old is New Again

The most recent moment in the history of digital distribution has been shaped by the fusion of old and new media, particularly in the forms of digital music and video onto the Internet. Peer-to-peer networking (such as Napster and a host of Napster-like clones) have provided a revolutionary method of file sharing, taking full advantage of the decentralized nature of the Internet's packet-switching nature. The transformation of analog content (film, images and music) into a digital medium is more than just a translation of wave forms into 1s and 0s, it represents a transformation in how information is communicated and distributed as much as how content is managed. It also means that a history of strict regulation and control of old media is continually being challenged by the ethical framework of new media technology.[18]

It is not surprising, given the history of digital distribution, to discover then that a large segment of the public perceives content on the Internet as free, even while they acknowledge that it may be copyrighted and regulated in its distribution in other media. The <u>ethos</u> of new media has fostered a belief that while content can be owned, controlled, and regulated, <u>distribution</u> cannot (and should not) be. As a result, the value of the Internet is not found in the information it provides, but in the <u>way</u> in which it provides that information. File sharing (with products such as Napster) differs little in form from the traditional ways in which music and even film have been trade among friends (e.g., tape recording a CD or album for a friend). The difference rests with the degree to which the network of distribution has become public and global.

The problem that old media faces is that the ethic which has always driven new media is one which celebrates the idea of sharing information. If something <u>can</u> be shared, this ethic dictates, then it <u>should</u> be shared. Hardware (the physical computer itself) as a set of atoms is what demands purchase. Information, however, as merely the arrangement of 1s and 0s, flows too freely and too easily to be regulated. Whether than information is an email sent to a friend or full motion video of the latest blockbuster release, it is all reducible to the arrangement of 1s and 0s to be interpreted by a machine.

There is a clear sense too that this shifting from physical to digital media is forcing the industry to rethink the process of production. New forms of copy protection, invariably, lead to new cracking techniques, forcing the industry to continually innovate. Digital "wrappers" which require payment to allow a user to listen to a song, also allow digital music to be transmitted via Internet, requiring each user along the line to pay a similar fee. Microsoft's latest operating system, XP, require users to register the OS with Microsoft based on their computer's hardware configuration, insuring that the software can only be installed on one machine.

Not all examples of the shift to digital distribution have been so adversarial between end users and industry. In one case, Miramax's script for a film *Takedown*, the story of the capture and arrest of Kevin Mitnick, was pirated and circulated among hackers on the Internet. Because it was a computer file, it was easily distributed and was generally sent in encrypted form, minimizing the risk of liability to those who possessed copies of it. The script, which had serious and even potentially libelous errors in it, was the subject of a number of protests and letter writing campaigns. As a result, Miramax ended up significantly altering not only the script, the entire tenor of the film to be much more sympathetic to Mitnick's point of view. The producers even went so far as to hire on acquaintances of Mitnick to supervise production and attend to factual accuracy.

As entertainment media such as music, videos, films, and books, which have undergone the transformation from analog to digital, begin to outstrip the physical media which were once necessary for their distribution, old media content providers are going to find it increasingly difficult to stake claims of ownership of networked bits. Not withstanding recent legal rulings, which have in all cases upheld the rights of content owners, it is going to become increasingly difficult to enforce intellectual property rights as the more fully decentralized nature of the Internet plays a more prominent part in media distribution and as long as the tradition of the hacker ethic, demanding the free flow of digital information, subsists within the network and the networked community.

Notes

1. One example worth noting is the Communication Decency Act (CDA) of 1996, which attempted to control indecent content on the Internet by subjecting it to the same standards used to control broadcast. The CDA was, in fact, an amendment to the 1934 Telecommunications Act that established, among other things, the FCC. The CDA was ruled unconstitutional based primarily on its uniqueness as a communication medium and was found to merit, because of its nature, the highest standards of protection.

2. Much of the tension stems from a number of high profile contestations of the legitimacy of intellectual property law in cyberspace. Most notably, John Perry Barlow's essay 'Selling Wine Without Bottles: The Economy of Mind on the Global Net' published in Wired under the title 'The Economy of Ideas: A framework for rethinking patents and copyrights in the Digital Age (Everything you know about intellectual property is wrong)', Vol. 2, No. 03, March 1994, pp. 84-90.

3. Microsoft, 'What is software piracy?' Windows ME, 2000.

4. This issue is taken up by Mark Poster in his book The Second Media Age (London: Blackwell, 1995), particularly around notions of the ways in which consumption and production are blurred by new media.

5. I used these terms not to suggest a binary distinction, but rather as a means to discuss the only the means of reproducibility. I do not want to suggest

that computers are incapable of producing art or being used in an artistic fashion any more than I want to suggest that art functions without codes or rules. Instead, I use the terms art and code to discuss the primary means by which these different media are rendered reproducible.

6. Plato, Sophist, 236b.

7. Gilles Deleuze, Difference & Repetition, trans. Paul Patton, (New York: Columbia University Press, 1995), pp. 38-39.

8. Walter Benjamin, 'The Work of Art in the Age of Mechanical Reproduction', in Hannah Arendt (ed.) Illuminations, trans. Harry Zohn, (New York: Schocken Books, 1969), pp. 217-252.

9. Ibid., p. 220.

10. Ibid., p. 221.

11. Microsoft, 'End User Licence Agreement', Windows ME, 2000.

12. Microsoft, 'What is software piracy?'

13. Steven Levy, Hacker's: Heros of the Computer Revolution (New York: Bantam, 1984), p. 40.

14. Nicholas Negroponte, Being Digital (New York: Viking, 1995), pp. 11-17.

15. Bruce Sterling, The Hacker Crackdown: Law and Disorder on the Electronic Frontier (New York: Bantam, 1993), p. 66.

16. PWA, .nfo file, 21November 1996.

17. RiSC, .nfo file, 21November 1996.

18. For a very sophisticated discussion about the relationship between the Internet and regulation, see Lawrence Lessig's Code and Other Laws of Cyberspace (New York: Basic Books, 1999). Lessig argues that code, as the self-inscribed architecture of the Internet, creates an internal system of regulation that will ultimately be the means by which intellectual property rights on the Internet are negotiated.

Chapter Two: Hacking the Body: Code, Performance and Corporeality

Narratives of technology are routinely given to binary distinctions. Consider, for example, the impulse to construct narratives of technological innovation as either utopian or dystopian. Every celebration of the emergence of a new technology has been accompanied by darker vision, usually spelling the end of civilization as we know it. Television, for example, was heralded as a technological breakthrough, while at the same time, Congress convened in the 1950s to hold hearings to determine if television caused juvenile delinquency.[1] Later, throughout the 1990s, the rise and growth of the Internet generated promises of world peace, the dissolution of international boundaries, and the improvement of race relations, while at the same time triggering fears about user addiction and depression—and, most of all, about the figure of the (typically male, teenage) hacker.[2]

Theoretical inquiry also reinforces these concerns: from Martin Heidegger's 1954 essay, "The Question Concerning Technology," which considers technology's "greatest danger" in science as well as its "saving power" in art. to the debates over technological determinism and constructivism, scholarly considerations of technology seem to be continually reducible to pairs of terms and to the negotiation between them.[3] One of the reasons for that, I want to suggest, has to do with the fact that technology always performs a function of mediation, generally between two poles, and our discourse and even our theory tends to recapitulate that tension.

Hence, throughout my inquiry, I will position, or perhaps reposition, the idea of technology between two terms that often function in opposition to one another: code and performance. In doing so, I want to suspend the meaning of technology between the two, to look at technology not as a product but as a process that generates meaning through the negotiation of these two conceptual and material elements. Of course, it would in no way be fruitful to attenuate the force of this opposition. But, I do think there is some value in examining what the tension between the two entails—in examining the poles not as mutually exclusive options, but rather as anchoring points, nodal points, if you will, for the broader discussion of the meaning of technology itself.

Proceeding in this manner, I will demonstrate that technology embodies aspects of both code and performance and we can gain a better of understanding of how technology functions that by interrogating the ways in which technology both utilizes and problematizes each of these dynamic factors. In particular I want to examine the ways in which the development and uses of code, particularly in the cases of open source development and encryption can be better understood as dynamics of code and performance within the context of hacking subcultures of development and subcultural resistance.

Most directly, I locate much of my analysis in the relationship between bodies and codes. While it is easy to understand the ways in which performance can be understood corporeally, code is a much different matter. As a system of abstraction, code tends to be removed from the body. Within the various contexts of code and performance, this essay is an effort to rethink the boundaries between the natural and artificial, between the real and virtual, and, ultimately, between code and performance. The body, I will argue, is the site where the logics of such binary division can no longer sustain themselves as separate and the point at which we can begin to understand the meaning of hacking as resistance.

Code

Defining one's terms is always a dangerous business, so let me begin by saying that I am doing so provisionally and somewhat advisedly. That said, the definitions I propose, with respect to the facts and phenomenon of code and performance are themselves designed to be *useful*, rather than definitive. I begin with the crux of the matter at hand, with the lifeblood of technology and hacking in the late second half of the 20th century: code. By code, I mean to refer to a system of regulation, a regime, which is both structured and *structuring*, which is to say that at its base, the primary function of code is *normative*. This is a vital aspect of code, one upon which my subsequent arguments will pivot and one that they will push to its limits as well.

I take as a grounding assumption the idea that, as Larry Lessig asserts, "code regulates," and that in the case of the Internet, the medium that occupies a great deal of his attention, we must understand "how the software and hardware that make cyberspace *what* it is regulate cyberspace *as* it is."[4] In other words, the machinery and code that constitute the substance of cyberspace condition not only what it means, but also how it means. Accordingly, even the most basic functions of computer code, loops and branches, variable assignment and calculation (functions shared by all computer programming languages), describe the basic parameters of what is both possible as well as what is acceptable.

Lessig also argues that code, specifically computer code, shares its normative function with law.[5] As law, in the general sense of the word, code functions abstractly, always moving from the particular to the general in an effort to define the boundaries of normalcy. The premise of law itself is grounded in the regulation of social function, abstracted from particular instances of behavior or events which are deemed unacceptable or injurious to the community.[6]

Moreover, as a general concept, code goes well beyond the confines of perl, C++ or the court system. Code is endemic to any system of writing. In broad terms, code is grounded in its own repetition, its iterability. Code only functions when it appears to repeat without difference, even as the situations, contexts, and evens to which it is applied vary in their specificity. Writing is transformed from the general structure of gestural codes and iteration into code at the moment when pen touches paper and ink marks the page—in the instance when "we simply do not know what our writing does."[7] Writing becomes code when it produces a surface effect that appears so completely unrelated to the underlying processes of creation that those processes altogether disappear from the finished product.

But, code becomes wedded to technology at the moment it begins to effect writing by in some way making it permanent, in those instances when, as Friedrich Kittler has noted, "modern media technologies in general, have been explicitly contrived in order to evade all perception."[8] In such circumstances, code is the means by which writing becomes alien to us, the means by which writing is transformed, and ultimately, in this day and age, reduced to a measure of differences in voltage registered by silicon chips. It is a process that is repeated throughout the history of technology in what David Bolter and Richard Grusin have called the dual logic of remediation, the process by which technological sophistication has, at base, the goal of rendering itself invisible even as its presence in the process of mediating communication increases.[9]

Think, for example, of the annoyance that occurs when we are reminded of the fact that we are speaking on the phone as a result of line noise or more disturbingly about a "reality" TV program like Survivor: the excesses of technology making the show possible stand in stark contrast to the abject denial of modern technology to the participants of the show, 16 cast-always forced to live off the land.

Thus, code has a second, equally important quality: the ability to make itself disappear, to vanish. Now, this insight should come as little surprise to us. The most effective systems of regulation and social control are always presented to us as pre-existing, natural, or essential. Thus, we might supplement Lessig's insight with the notion that the disappearance of code is an essential part of its functioning. That said, I want to foreground this notion of code as an abstract system that resists any attempt at particularization and that, at every opportunity, attempts to erase the means and manner of its functioning, leaving only the effect of code (a kind of surface effect) which makes the underlying processes of appearance indecipherable.

Consider the relationship between source code and software in most retail computer products. The software bears no trace of the source code that produced it. Nowhere in Microsoft Word or Windows XP can you find the source code, the programming that gave rise to the software. Law is so deeply interconnected with this phenomenon that when such code does emerge the law (and law enforcement) is immediately invoked. So, for example, when portions of Microsoft's source code for Windows 2000 and NT were posted anonymously on the Internet in February of 2004, Microsoft's response illustrated these assertions convincingly:

> Microsoft continues to work closely with the U.S. Federal Bureau of Investigation and other law enforcement authorities on this matter. Microsoft source code is both copyrighted and protected as a trade secret. As such, it is illegal to post it, make it available to others, download it or use it. Microsoft will take all appropriate legal actions to protect its intellectual property. These actions include communicating both directly and indirectly with those who possess or seek to possess, post, download or share the illegally disclosed source code[10]

The notions of secrecy emerge as almost the very definition of the code. It is not only illegal to post or distribute it; it is illegal to possess it and use it. The juxtaposition is striking. In an industry where the goal is to have each and every computer running your software, the very success of that enterprise is dependent upon making the source, literally the code itself, invisible, inaccessible and, ultimately, bringing the full force of law to bear on those who seek to violate that principle.[11]

Performance

Performance, however, is different. It is the opposite of code. In Peggy Phelan's discussion of the ontology of performance, she writes: "Performance's only life is in the present. Performance cannot be saved, recorded, documented, or otherwise participate in the circulation of representations of representations: once it does so, it becomes something other than performance."[12]
Performance, at least in the sense that I want to use the word, deals with that which is essentially *unrepeatable*. Of course, both code and performance repeat (or in the strictest sense of the word fail to), but what differentiates them is their direction. The goal of code is infinite repetition without difference, while the goal of performance

is the production of an irreducible, unrepeatable event—a singularity. Of course, one without a tinge of the other would lapse into utter unintelligibility. Code, in order to be in any way useful, must have a certain "eventfulness" about it, while performance, in order to be understood, must engage in repetition, even if that repetition is, as Judith Butler suggests, "parodic" or "subversive"[13]

Performance is literally a re-presenting or even re-presencing. It generates its authenticity not from repetition *per se*, but rather as repetition *with difference*. It is a calling attention to itself which questions, subverts, or even just calls attention to that which preceded it. It is everything that Plato abhorred about poetry, art, and music. Performance is the copy that strives not to be the same, but to be different. A *simulacrum*, an image without resemblance.[14] That is not only what gives it force, but also what, ironically, gives it authenticity. Accordingly, while the value of code rests in its normativity, the value of performance is found in transgression.

In groups like the Homebrew Computer Club, we find the origins of the "open source" software movement, which is grounded in the belief that all software should be distributed along with the code that produced it, making replication and, more importantly, alteration possible. But it is more than a guiding principle, for many programmers it rises to the level of an ethic, even a moral imperative.

It is also the site of primary antagonism between open source culture and Microsoft. In 1976, Micro-Soft was a small start-up in Albuquerque, New Mexico, selling the computer language BASIC (a language released into the public domain by a pair of Dartmouth programmers) to a handful of enthusiasts who subsequently reproduced and distributed the language to club members. Alarmed that users were trading code, rather than paying for it, a young William Henry Gates III published "An Open Letter to Hobbyists" in which he re-characterized the culture in dramatic terms. "As the majority of hobbyists must be aware," Gates wrote, "most of you steal your software. Hardware must be paid for, but software is something to share . . . the thing you do is theft." That antagonism still animates the open source/Microsoft split that is very much alive today, marking Open Source as transgressive in terms of capital, and, as importantly, in terms of the law. Open source is marked as a system of *unregulated* distribution and re-distribution that undermines the normative function of code and in doing so demonstrates the ways in which law and capital are complicit with each other.

By way of contrast, open source is performative and transgressive. This performative, transgressive dimension manifests itself in what I am calling the "culture of code." It is a space of repetition with difference and a space of questioning. It is the space in which the meanings, especially the cultural meanings of technology, are contested. It is the space that provides for acts of resistance within the very fabric of technology itself. These spaces or cultures have a very basic function in subverting the normative functioning of code. Accordingly, the traditional "hacker" often has little patience for authority, prefers decentralization to centralized systems, and uses words like "freedom," "art," "beauty" and "elegance" to describe the act of programming. In turn, across the board, while the product of code, software, reflects a highly regulated structure, the production of code is better described as performative. Often, we find that while programmers produce code, designed to be reproduced endlessly without difference, their own styles tend to be highly idiosyncratic, individualistic, and particular. This is especially true of open source programmers, such as Larry Wall, Richard Stallman, and Eric Raymond who are self-described "evangelists" and who describe the software they produce as "subversive."[15]

The more performative the act of coding (as is the case with open source software), the more likely it is that the process and the code itself will be rendered visible. Performance, like code, also disappears, but in doing so, it leaves a mark, not of its product (e.g. software) but of its process. Performance literally disappears into memory, but in doing so remains vivid and present. As such, a goal of performance is to disappear (as it must as an event bound in time and space) but in the process, it render itself opaque, lasting and permanent. Nowhere can this be seen more clearly that in the production of open source software. As the name suggests, open source software requires that not only their distribution, but also all subsequent distributions, include the source code that generates the software. The resulting product is considered "free software," not in terms of its price (you can charge for open source, though many don't), but in terms of its character. The user is "free" to change, alter, develop and re-use aspects of the code as they see fit. It is a transmission not only of the code, but also of the culture that generated the code in the first place. It is that aspect, the quality of "freedom" which cannot be in any way documented, but rather can only be performed through the process of distribution and re-distribution. It is that process that disrupts the smooth flow of production from code to product by re-presenting (presencing) the code at every turn. It is that performance which is performing the cultural, transgressive, and subversive aspects of the culture of code that is embodied in it.

Performance as the Disruption of Code

To say that the Open Source movement has disrupted traditional notions of software development and distribution would be an understatement. Not surprisingly, one of the venues in which the battle between code and performance is being fought is within the legal system itself. Brought to a head by issues of digital rights management, open source advocate's demands for "free" software and code has so fundamentally disrupted the software industry that Microsoft created a strategy designed to combat open source. In November of 1998, two documents were leaked from Microsoft headquarters, the so-called "Halloween Documents."[16] Although many issues were identified as "threats" to Microsoft (including the fact that OSS [Open Source Software] tends to be superior in quality), the primary issue was not code, but culture: "The ability of the OSS process to collect and harness the collective IQ of thousands of individuals across the Internet is simply amazing. More importantly, OSS evangelization scales with the size of the Internet much faster than our own evangelization efforts appear to scale." Community, culture, and evangelism are the motors driving both Microsoft and open source.

The Halloween documents also illustrate the failure of Microsoft's standard marketing technique, FUD, which stands for "Fear, Uncertainty, Doubt," a marketing technique deployed to scare consumers by questioning the permanence, reliability and compatibility of competitive products. They even included claims that Microsoft had implanted harmless "error" or warning messages in their operating system, which would alert users to the fact that certain programs were not "Microsoft tested" for compatibility. Nevertheless, Open Source was, by Microsoft's own testing, immune to FUD tactics and in order to combat the spread of open source software,

Microsoft acknowledged the necessity of retreating back to code and law. The only effective way to combat Linux was through patent and copyright. There were certain ideas and concepts Microsoft believed they could make legal claim to own, and thereby invoke proprietary ownership of certain software.

Code, Performance and the Body

Before moving to an extended example which addresses these phenomena more directly, I want to consider a third aspect that technology affects: corporeality. Code, in its most general sense, regulates and defines not by attaching itself to particular bodies, but instead by abstracting notions that appear to apply to *all* bodies. There is, and can never be, a particular "coded" body. To produce one would be to already raise an objection to the very abstracted, ideally universal conception of code. The "coded body" therefore cannot exist without disrupting the very notion of code itself. To produce such a body, in other words, would, *de facto*, mark the singular coded body as itself different—as a reiteration and realization of with at least one unique sign of substantive rearticulation. The incarnate, coded body, therefore, can only exist as a condition of its own impossibility. If it exists at all, it must exist in an abstract state, one which can never be realized or made flesh. Doing so, rendering the abstract material, would be to expose material or corporeal *absence* as the essential motor which drives the process of cultural, coded production.

Performance, on the other hand, *is* the body. It needs bodies to make itself real. It is the material and the flesh, the substance of being. It is the exception rendered visible which challenges not only the abstract category of code against which it is compared, but also the entire regime of comparison. In removing the essence, it removes the very possibility of appearance.

In what follows, I want to make concrete the theses I have just advanced. In doing so, I want to explore hacking as a context in which the tensions between code and performance provide for us a clear understanding of how technology both functions and fails to function as a system of cultural mediation between normativity and transgression, between repetition and the unrepeatable, and between abstract incorporation and corporeal materiality. In doing so, I want to make the case for a performative model of technology that understands and exploits the means of transgression that technology affords us through the thematic of performance. In doing so, I want to examine the various ways in which state control of encryption and hacker resistance to that control can helps us better understand the relationships among code, performance, the body and technology.

ITAR: Bodies, Codes and the State

My illustration deals with encryption. Encryption, as the process of encoding and enciphering text, appears on its surface as act of pure code. On the surface, it appears to have nothing to do with the body. Recent algorithms, in particular the RSA algorithm (named for its inventors, Ron Rivest, Adi Shamir and Len Adleman) have reduced encryption almost entirely to the language of mathematics. As a system of public key encryption, RSA provides a mathematically secure system that creates problems that are unsolvable without a key. Such public key encryption, which has become the standard for most Internet and government work, relies on mathematics no more complicated than basic prime number theory, algebra and modulus arithmetic. The principle of modern encryption is simple: one finds a mathematical problem that is simple to pose and impossible to solve without a key. Modern encryption utilizes even more sophisticated "one way" functions, allowing a code to have two different keys, one that locks and one that unlocks, neither of which can be derived from the other. The triumph of modern encryption is the removal of the body, rendering encryption an act of *pure* code.

Early efforts at encryption required a secret to be shared, passed from one body to another. It was predicated on both parties knowing the same secret, but our latest efforts remove even that problem. No one needs to know your code, your secret or your password, not even the computer. The history of encryption, however, tells a different story. In telling it, we begin not with mathematics or computers, nor with the codebreakers of the Second World War, or even as some have suggested with Julius Caesar scrambling messages to troop commanders in Rome.

The history of encryption begins with the last book of Oedipus trilogy, *Oedipus at Colonus*, where three discrete elements merge within the first narrative of encryption. Oedipus, fearing his grave would be defiled, extracts a promise from the King of Athens, to keep the place of his burial (literally his crypt) secret. In exchange, Oedipus offers Athens perpetual safety from its enemies as long as the secret is maintained.[17] These three components comprise the base elements of the technology of encryption: a code, the body and the protection of the state.

It also marks the body as the disruptive element which continually threatens not only the normative structure of code, but the very security of the state itself, especially when that body, like Oedipus's is a transgressive one, marked by its violation of code and its transgression of the law.

Of course, the more things stay the same, the more they invite performative subversion. In some 20th century contexts, for example, in contrast to the State, which utilizes encryption to keep secrets and guarantee security, a new generation, of self-proclaimed *cypherpunks* has found encryption to be an invaluable tool for subverting a chief aim of government: surveillance. For cypherpunks, encryption's killer application was not found in secrecy, but rather in privacy, in the possibility of selectively revealing yourself to the world. Anonymity and privacy became the chief tools in the battle against an increasing mechanized and panoptical security state. Code has once again found a way to become transgressive. With the development of RSA (the code I mentioned at the outset of this subsection) and a free, open source, application of it called PGP (Pretty Good Privacy), encryption was seen as a viable means for protecting the rights and freedoms for a digital generation.

But the State is never easily subverted and, as *cypherpunks* practiced their performative acts of resistance, the government retaliated with legislation restricting encryption. In 1991, Congress attempted to pass legislation mandating that all forms of encryption contain a "back door" allowing governmental access to coded civilian documents. It failed to pass into law, but was the impetus for Phil Zimmerman to develop and write PGP for public distribution. The government responded, in turn, by claiming patent infringement and, later, by classifying the RSA algorithm under ITAR (International Tarriff on Arms Restrictions). In doing so, for the first time, 3 lines of computer code become officially classified as a munition. Exporting it without official sanction from the State Department was no small crime. If convicted of exporting RSA, PGP or any other encryption program, one could face a million dollar fine and 10 years in jail for each export violation. The extremity of US policy was comparable to only five other nations, Russia, Iran, Iraq, China and France.

In an effort to fuse law and code, the State Department's efforts create a climate in which the export and distribution of code (and in some cases products) was heavily regulated based strictly on a system of classification and protection. RSA, in the new classification, deployed the tropes of "national security" as well as a number of other social and cultural threats, including organized crime, drug trafficking, terrorism, and pedophilia. In response, hackers began a movement to display the code wherever possible. The event which actually led to the compression of the RSA algorithm into 3 lines of inscrutable code, later refined by Adam Back into the following:
print

```
pack"C*",split/\D+/,`echo"16iII*o\U@{$/=$z;[(pop,pop,un
pack"H*",<>)]}\EsMsKsN0[lN*1lK[d2%Sa2/d0<X+d*lML
a^*lN%0]dsXx++lMlN/dsM0<J]dsJxp"|dc`
```

With these two lines of perl code, the guts of the RSA algorithm became at once completely obfuscated (even the most wily perl hacker would have difficulty deciphering Back's code), while at the same time rendering it, on a different level, completely and totally accessible. Reduced to two lines of code, it could be reproduced and disseminated not as code *per se*, but rather as a statement about the reproduction and dissemination of code. In other words, Back's code transformed the RSA algorithm from code into performance. In its more verbose version, three lines long, the code first appeared on T-Shirt.

The significance of this transformation is that it both rendered the code visible (and hence performative and disruptive) while at the same time completely neutralizing it. The T-shirt couldn't encrypt anything, its only value was in the *visibility* of the code, not in the function of the code itself.

Later, it emerged on a web page, the ITAR Civil Disobedience, alternatively titled the "International Arms Trafficker Training Page" which billed itself as an act of civil disobedience and in doing so, allowed anyone to export banned crypto anonymously. Additionally, there were boxes to check to add your name to a list of "known arms traffickers" or to send a letter to the President declaring your act of civil disobedience.

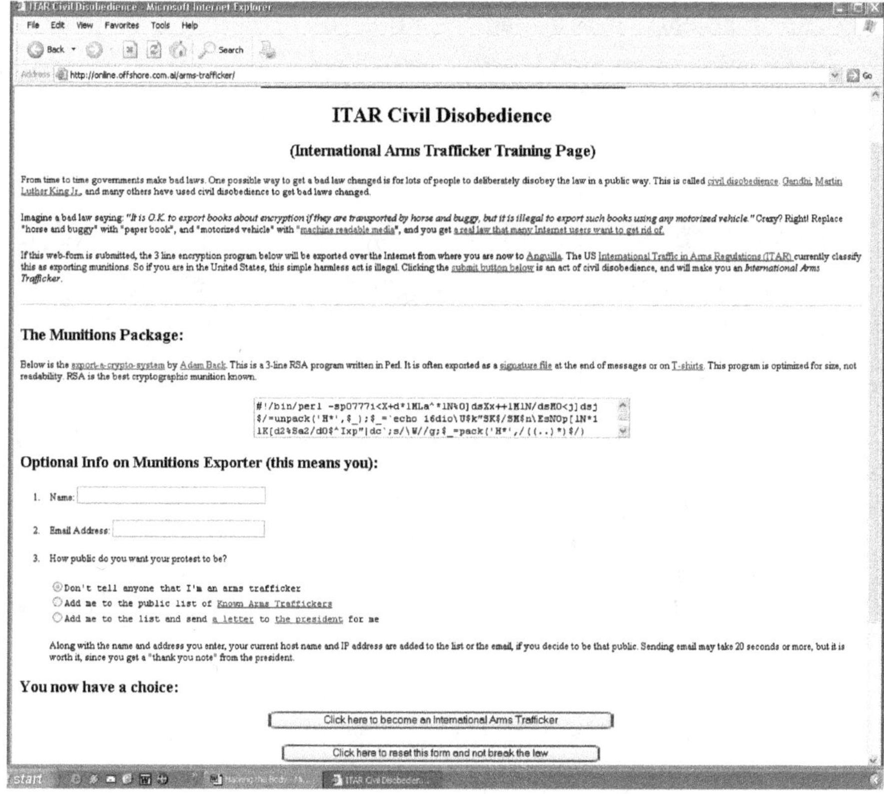

ITAR Civil Disobedience

(International Arms Trafficker Training Page)

From time to time governments make bad laws. One possible way to get a bad law changed is for lots of people to deliberately disobey the law in a public way. This is called civil disobedience. Gandhi, Martin Luther King Jr., and many others have used civil disobedience to get bad laws changed.

Imagine a bad law saying: *"It is O.K. to export books about encryption if they are transported by horse and buggy, but it is illegal to export such books using any motorized vehicle."* Crazy? Right! Replace "horse and buggy" with "paper book", and "motorized vehicle" with "machine readable media", and you get a real law that many Internet users want to get rid of.

If this web-form is submitted, the 3 line encryption program below will be exported over the Internet from where you are now to Anguilla. The US International Traffic in Arms Regulations (ITAR) currently classify this as exporting munitions. So if you are in the United States, this simple harmless act is illegal. Clicking the submit button below is an act of civil disobedience, and will make you an *International Arms Trafficker*.

The Munitions Package:

Below is the export-a-crypto-system by Adam Back. This is a 3-line RSA program written in Perl. It is often exported as a signature file at the end of messages or on T-shirts. This program is optimized for size, not readability. RSA is the best cryptographic munition known.

```
#!/bin/perl -sp0777i<X+d*lMLa^*lN%0]dsXx++lMlN/dsM0<j]dsj
$/=unpack('H*',$_);$_=`echo 16dio\U$k"SK$/SM$n\EsN0p[lN*1
lK[d2%Sa2/d0$^Ixp"|dc`;s/\W//g;$_=pack('H*',/((..)*)$/)
```

Optional Info on Munitions Exporter (this means you):

1. Name:
2. Email Address:
3. How public do you want your protest to be?

Don't tell anyone that I'm an arms trafficker
Add me to the public list of Known Arms Traffickers
Add me to the list and send a letter to the president for me

Along with the name and address you enter, your current host name and IP address are added to the list or the email, if you decide to be that public. Sending email may take 20 seconds or more, but it is worth it, since you get a "thank you note" from the president.

You now have a choice:

Click here to become an International Arms Trafficker

Click here to reset this form and not break the law

Ultimately, however the performative acts of resistance would be taken to what I will argue is their logical extreme: to the fusion of bodies and code, which marks the body as irreducibly transgressive. In this same vein, the cypherpunk movement itself clearly recognized and powerfully deployed corporeality as means to resist through the medium of tattooing. One of the first cypherpunks to acquire an RSA tattoo did so as an effort to raise public awareness about issues of encryption. Thus, in this instance and in others like it, the literal embodiment of code, announced its own performance not as an act of encryption, but as an act of resistance. It therefore was the ultimate statement of a "culture of code," allowing the body to speak the language of code, rending the code both transgressive and impotent. It also had the effect of rendering a number of publications complicit in arms trafficking, as pictures of the image appeared in Japanese, French and UK publication as well as the New York Times

.

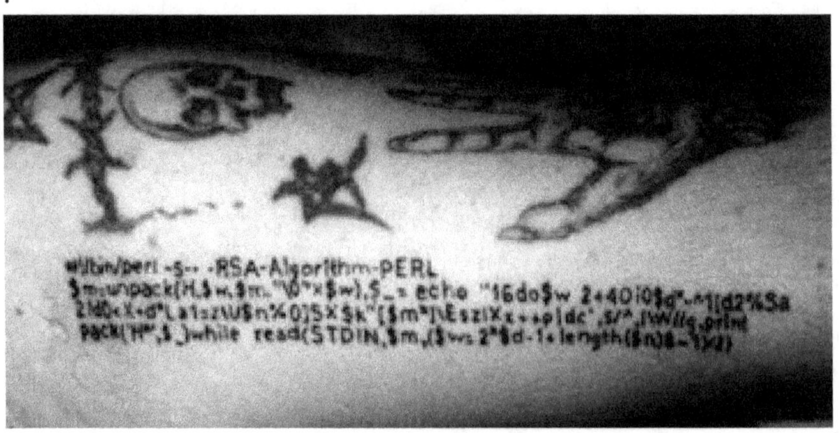

In doing so, the performative act of creating a coded body, defies precisely what the law allows. In other words, by forfeiting function and making the code appear where it cannot, these marked bodies became transgressive and disruptive in a way that law simply cannot reconcile.

A second celebrated RSA tattoo, that of cypherpunk known as CancerOmega, is even more explicitly subversive, bearing the warning “This Man is Classified as a Munition.”

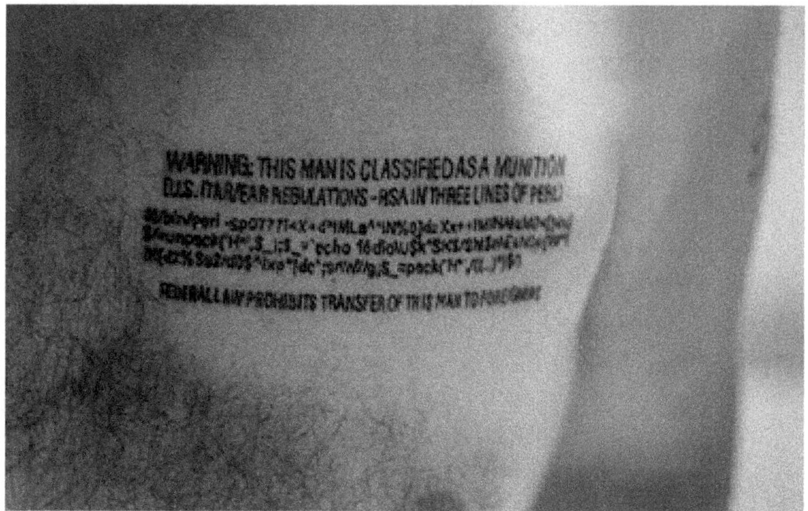

By marking the source and origin of the law’s code (the classification of code as a munition), CancerOmega has exposed something fundamentally wrong with the notion of law and code. This is code pushed to its limit. And if we are to conclude that the statement “This man is classified as a munition” is at once technically correct as a matter of code, law and regulation, we are forced at the same moment to confront the absurdity of such a statement through performance. The law, when confronted with the coded body, loses its force as Law: it ceases to function as an abstract principle of normative regulation and, instead, becomes a parody or itself.

In addition, speculative newsgroup and listserv discussions appeared regarding other ways that such performances might further subvert the law. When foreign students, particularly those studying computer science, learned of the export regulation attached to the

RSA algorithm, students considered tattooing their bodies with the code to prevent them from being deported when their student visas expired. In a post to alt.callahans (08/28/98), Steve Brinich suggested the following:

> an illegal immigrant might get an RSA-in-perl tattoo in order to prevent deportation, since it's illegal to export and US law won't allow the government to subject people to involuntary surgery.

Similar discussions proliferated on lists ranging from a cypherpunk newsgroup to rec.arts.bodyart to any number of perl newsgroups and mailing lists. One exchange (on comp.org.eff.talk) highlighted still more performative acts. One poster's suggestion for an RSA tattoo included additional subversions:

> Post: You know, my wife's a tattoo artist, and I've been seriously toying with the idea of getting the RSA algorithm tattooed somewhere on my body. Might be interesting to call a press conference and then try to board an international flight. What are they going to do- tell me I can no longer leave the country, as I'm now classified as a "munition" and thus prohibited from export?
> Such a suggestion drew the following response:
> Reply: Might I suggest the best place for this would be on your posteriour [sic]? That way, when you had to "present evidence", you would get to moon the court and law enforcement officals[sic]!! What better statement could one make about the whole situation?!?!?[18]

Here, again we return to my argument that the performative act of transforming codes into bodies (and also bodies into code) potentially disrupts the entire system of regulation and normativity that allows code itself to function. As I have demonstrated, such disruption is generated by making the code appear, and that disruption is, in the strictest sense of the word, a performative act.

Conclusions: Code and/as Performance

This essay has endeavored to chart an understanding of technology that is opened up through performance. That opening is not merely one that mediates technology, but is the space in which resistance and transgression both become possible. As I hope I have argued convincingly, performance is no more a-technological than technology is non-performative. As a consequence, the *politics* of technology have a great deal at stake in their performative dimension insofar as the relationship between performance and technology can perhaps be best described as one of intervention. Importantly, while that intervention does not necessarily require one to write code (although it in no way excludes that capability either), it does require one to be able to negotiate code, to translate code, and ultimately be able to re-deploy it in ways which challenge its normative functioning.

What this literal hacking of the body demonstrates is perhaps the most deeply held and most important element of the hacker ethos: that resistance is always about finding alternative and profound means to transgress. What characterizes hacktivism as a unique form of resistance has more to do with process than it does with effects or outcomes. Hacktivism is, at base, a deconstructive endeavor, which seeks not to undermine any single idea or expression, but, instead, which seeks to undermine the very foundations and logics upon which such expressions are built.

References

Benjamin, W. (1975) “Critique of Violence,” *Reflections. Essays. Aphorisms, Autobiographical Writings.* New York and London: Harcourt Brace Jovanovici.

Bolter, J. D. and Grusin, R. (1999). *Remediation: Understanding New Media*, Cambridge: MIT Press.

Butler, J (1989) *Gender Trouble*, New York: Routledge.

Deleuze, G. (1990) “The Simulacrum and Ancient Philosophy” in *The Logic of Sense*, trans. Mark Lester, New York: Columbia University Press.

Derrida, J. (1992) “Force of Law: 'The Mystical Foundation of Authority' ", in *Deconstruction and the Possibility of Justice*, edited by D. Cornell and M. Rosenfeld, New York: Routledge.

Heidegger, M (1993/1953) “The Question Concerning Technology,” *Basic Writings*, New York: Dimensions.

Kittler, F. (1995) “There is no software” CTHOERY, 10/18/1995 [available at http://www.ctheory.net/, last accessed September 12, 2004]

Lessig, L. (1999) *Code and Other Laws of Cyberspace*. New York: Basic Books.

Lessig, L. (2003). *The Future of Ideas: The Fate of the Commons in a Connected World*, New York: Random House.

Lessig, L. (2004) *Free Culture: How Big Media Uses Technology and the Law to Lock Down Culture and Control Creativity*, New York: Penguin.

Microsoft . (2004) “Statement from Microsoft Regarding Illegal Posting of Windows Source Code, Feb. 20, 2004” .Microsoft, Redmond, WA.

Nissenbaum, H. (2004) "Hackers and the Contested Ontology of Cyberspace" *New Media and Society*, 6:2: 195-217.

Open Source Initiative. (2005). Available at http://www.opensource.org/halloween/, last accessed March 17, 2005.

Phelan, P (1992) *Unmarked: The Politics of Performance*, New York: Routledge.

Raymond, E (2001) *The Cathedral & the Bazaar*. Cambridge: O'Reilly

Sophocles (2004), *Oedipus at Colonus*, trans. By E. Grennan and R. Kitzinger, Oxford: Oxford University Press.

Sturken, M. and Thomas, D. (2004) "Technological Visions and the Rhetoric of the New," in M. Sturken, D. Thomas, and S.J. Ball-Rokeach, *Technological Visions: The Hopes and Fears that Shape New Technologies.* Philadelphia: Temple University Press.

Thomas, D. (2002) *Hacker Culture*, Minneapolis: University of Minnesota Press.

Notes

[1] Sturken and Thomas, pp. 6-16.
[2] For a discussion of the primarily male, teenage, suburban image of the hacker as "boy culture" see Thomas, *Hacker Culture*, especially pp. xii-xvii75-6. 158, 161, 206-212. For a particularly interesting examination of how the term hacker has changed to include the notion of these threats see Nissenbaum (2004).
[3] Heidegger (1993).
[4] Lessig, *Code*, p. 6.
[5] This theme animates much of Lessig's more recent work as well in *The Future of Ideas: The Fate of the Commons in a Connected World*, and more recently in *Free Culture*: *How Big Media Uses Technology and the Law to Lock Down Culture and Control Creativity*. Both can been seen as efforts to put this fundamental premise into action, especially in relation to reforming copyright and its effect of creativity.
[6] Ironically, these are also themselves moments of injury of violence. On this point see Benjamin's "Critique of Violence," and later, Jacques Derrida's "Force of Law: 'The Mystical Foundation of Authority"."
[7] Kittler (1995).
[8] Kittler (1995).
[9] Bolter and Grusin (1999).
[10] See Microsoft's press release, "Statement from Microsoft Regarding Illegal Posting of Windows Source Code."
[11] In many ways this can be seen as the origins of hacking as political action. See, for example, Thomas (2002), pp. 100-110 especially. Also of note is the fundamental transition that takes place in the late 1990s where hackers engage in direct political intervention by making code appear, rather than through acts of disruption or defacement. Although I would never claim that these two are mutually exclusive, the spirit of this distinction is an important one, which requires one to pay careful attention to the development of the subculture itself.

[12] Peggy Phelan (1992). In particular see chapter 7, "The ontology of performance: Representation without reproduction."
[13] Butler (1989), pp. 139; 142-149.
[14] For an in depth discussion of the relationship between simulacra and resemblance, see Deleuze (1990).
[15] Raymond (2001).
[16] Open Source Initiative (2005).
[17] Sophocles (2004), see especially lines1732-1740.
[18] Interestingly, the discussion was in response to Richard White's original post, considering the tattoo he eventually did have done. Posted 07/03/1995 to comp.org.eff.talk.

Chapter Three: KPK, Inc.: Race, Nation, and Emergent Culture in Online Games

Digital media and learning pose a set of central problems for research and educational practice. As these spaces become increasingly social and participatory, they form a complex social and cultural matrix which both draws from and contributes to discussions about social and cultural issues. Games, like many other new media, crossed a crucial threshold when they began to provide a social context for learning and interaction. Accordingly, it has become difficult, if not impossible, to limit our understanding of digital media and learning to the text or even the experience of particular new media technologies. What happens in the context around a game or web site can be as important as the content of the game or site itself.

Learning, then, occurs both through the experiences of interactivity in virtual spaces as much as it does in the broader context of social and cultural interaction for which these spaces serve as a shared reference point. Accordingly, one of the ways to understand the intersection of digital media and learning is to explore the practices that happen at the borders and boundaries of the media we examine (Thomas and Brown, 2006).

In the case of race and ethnicity, the issues are compounded by a variety of forces which shape our understanding of the world and our place in it. This essay is an effort to contextualize questions of race and ethnicity within game worlds and with gamer culture more broadly by examining a particular set of conflicts that emerged within the game Diablo II, shortly after its release in Korea in 2003. The conflict that arose illustrates one of the ways in which notions of race and ethnicity are constructed, defined, and contextualized in virtual spaces. It also provides some insight into how notions of privilege and investment in economic and social power relationships of the physical world are mimicked and intensified in the context of virtual spaces.

Race and Cyberspace

When one considers the complexities of race and ethnicity inside virtual worlds, the already difficult questions of what constitutes race are magnified. The temptation and tendency, as Beth Kolko has illustrated in the case of text based MUDs,[1] is to erase the category completely. As Kali Tal wrote in (and to the readers of) Wired, "I have long suspected that the much vaunted 'freedom' to shed the 'limiting' markers of race and gender on the Internet is illusory, and that in fact it masks a more disturbing phenomenon--the whitinizing of cyberspace." (Tal, Dec 17, 2001).

Both Kolko and Tal point to the disappearance of race as a signature of the whiteness of cyberspace. This essay examines the question of race and ethnicity from a distinctly different point of view, attempting to understand how race and ethnicity are, in fact marked by geography, language, and specific cultural practices within virtual worlds and multiplayer game spaces and how those markers might provide insight both into the way race functions both in cyberspace as well as in the cultural imagination of those who inhabit these worlds.

As George Lipsitz argues, race in American culture is often a matter of "possessive investment, " suggesting that those who have power both invest and are invested in the privileges that accompany whiteness is society. Lipsitz's goal is to "stress the relationship between whiteness and asset accumulation to connect attitudes to interests" and "to demonstrate that white supremacy is usually less a matter of direct, referential, and snarling contempt and more a system of protecting the privileges of whites by denying communities of color opportunities of asset accumulation and upward mobility."[2] Following Lipsitz, this essay addresses some of the ways in which cyberspace culture generally and gamer culture more specifically participate in cultures of racism through strategies of denial, refusal, and equivocation in an effort to maintain the status of white privilege. While racism exists in gaming culture in many explicit forms, some of which I will articulate below, it is important to understand the ways in which participating in discourses which deny racism are equally important and serve as a foundational discourse for learning in the context of digital media.

This essay, then, looks both at the forms of direct, hostile racism as well as the ways in which racist discourse often engages in acts of possessive investment in whiteness through strategies of disavowal. This essay provides a reading of the discourse of gamers confronting issues of racial, geographical and ethnic differences and it suggests some insight into the events that have triggered the rise of certain racist practices in one particular game. This discussion also aims to understand the process by which some American youth process notions of race and ethnicity in virtual spaces and the various entitlements and possessive investments they feel in relation to the virtual world of gameplay.[3]

While significant discussion about the impact of games has focused attention on violence and explicit content, very little attention has been paid to the more subtle forms of learning which occur in and around the spaces of games.[4] By broadening the scope and boundaries of games and play, we can begin to understand how various cultural forms and productions emerge that can help us better understand issues of race and ethnicity in games and culture.

Gamer Culture, Nation, and Race: KPK Inc and the "Korean Problem"

In the past decade, massively multiplayer online games (MMOGs) have emerged as a central form of entertainment in the digital age. MMOGs are fundamentally social environments where players create and name personal avatars, interact with other characters through gameplay and chat, and create and negotiate meaning in the game world through social, economic, and cultural exchanges and interactions. The games represent a fusion of the logics of text-based MUDs, which were heavily social in nature and allowed players to creatively construct identities through textual, literary production, and graphical adventure games, which provided for game interactivity in a primarily graphical non-modifiable context. This combination preserved the flexibility of complex identity play from MUDs with fast paced game play and interactivity more characteristic of adventure and first-person shooter games.

As a result, the blend of these two types of worlds has produced a new and arguably revolutionary framework for play, which is both social and individualistic, and grounded in game mechanics of player advancement, combat, and acquisition of items, treasure, and money. While much attention has been focused on traditional questions of media effects, for example the relationship between games and violence, little attention has been paid to the emerging intercultural dynamics of this new medium and its unforeseen consequences, especially regarding matters of race and ethnicity. Because MMOGs function as microcosms of larger social configurations, the ways they can be used, and the communities of practice that they spawn, can be complex and culturally rich.[5] Most importantly, because these games are spaces in which cultural meanings are both created and negotiated, these games are sites which helps us understand the ways in which the cultural imagination of players is being shaped around issues which they might not necessarily confront in their daily lives.

MMOGs have become a worldwide phenomenon, with extensive player bases throughout North and South America, Europe, and Asia. Because they are Internet based, players from different countries are able to interact and play together with relatively few problems. Consequently, it is not uncommon in these spaces to see players from England, Sweden, Germany, America, Canada, Korea and Japan all playing the same game on a single server. Among America gamers, such interactions may be their first, and in many cases only, interaction with citizens of other nations, which is quite significant in and of itself. In many cases, these interactions can be positive and fruitful learning experiences.[6] In others they can create and foster racist attitudes, which promote stereotyping. I am interested in examining some surprising and disturbing outcomes of transnational adoption within the context of one such MMOG, Diablo II and by examining the fallout of the game's adoption in Korea, especially in relation to player groups, such as KPK, Inc., a group formed outside of the game space with the expressed purpose of eliminating Korean players from the U.S. game space (KPK stands for "Korean Player Killers").

Diablo II

On June 28, 2000, Blizzard Entertainment released Diablo II, a follow-up to their enormously popular 1997 Diablo game. By July 17th, Diablo II had sold more than 1 million copies, making it the fastest selling video game at that time in PC history.[7] Within 6 months Diablo II had sold more than 2 million copies and had been named "Game of the Year" by a host of gaming magazines and websites, marking it as one of the most successful PC games of all time.

As Blizzard describes it, Diablo II is set in the context of an "eternal struggle to decide the fate of all Creation" which has "now come to the Mortal Realm . . . . After possessing the body of the hero who defeated him, Diablo resumes his nefarious scheme to shackle humanity into unholy slavery by joining forces with the other Prime Evils, Mephisto and Baal. Only you will be able to determine the outcome of this final encounter." The game is set in a "world of dark fantasy," where players play "one of five distinct character types, explore the world of Diablo II -- journey across distant lands, fight new villains, discover new treasures, and uncover ancient mysteries, all in the quest to stop the Lord of Terror, once and for all..."[8] Diablo II follows the lore of the earlier game and is set in a fictional world and time, which has no identifiable geographical or temporal connection to the present day world.

Players are able to select from one of five possible characters: Amazon, Sorceress, Barbarian, Paladin, and Necromancer. The Amazon and Sorceress characters are female, while the Barbarian, Paladin and Necromancer characters are male. Avatars, the player's graphical representation in the game, are not modifiable, resulting in significant attention being paid to naming conventions for the purposes of identity. Unlike more recent MMOGs which allow significant customization of the player's avatar, each character class in Diablo II was represented by a particular piece of game art, which meant that every player playing a barbarian, for example, would have an identical in game appearance. In that sense, the only way a player can differentiate him or herself from the other players in the game is through choosing their name.

The other point of differentiation was in the armor and weapons the characters used, which were significant parts of their characters identity and which determined, in large part, how successful the character was in the game. These items, which were treasure received from slaying monsters, were the primary resources for the game and were seen as barometers of a player's skill and status.

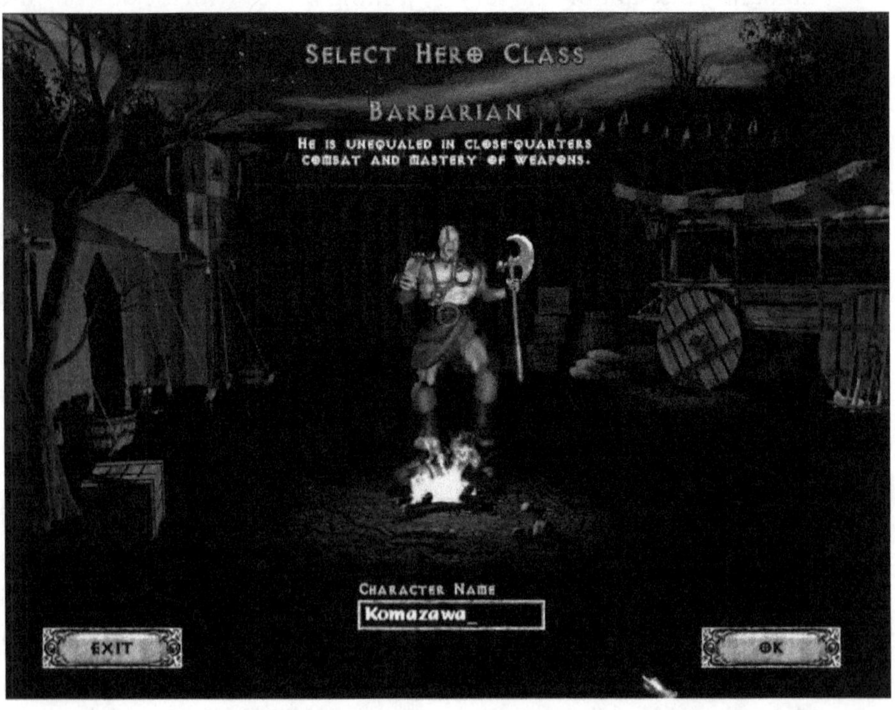

Character screen illustrating a class avatar for Barbarians

The game consists of five primary areas which each contain a number of "boss" monsters who randomly drop magical items that can be used to enhance your character or that can be traded to other characters in exchange for items or gold. Each of these five areas can be played at three different levels of difficulty (normal, nightmare, and hell), with the value of items increasing along with the levels of difficulty. As difficult levels increase, the monsters that players battle remain the same, but the monsters gain more hit points and resistances, making them exceptional adversaries that are much harder to kill.

While it was possible to play Diablo II as a single player game, its popularity (and profitable sales) was due in large part to Blizzard's free online gaming network, battle.net, which allowed anyone who purchased the game to play in an interactive multi-user environment. Anyone who purchased the game could connect to battle.net through the Internet and play cooperatively or competitively with other players who were also logged on.

A typical Diablo II game space, English UI

One of the primary goals of battle.net was to provide a setting for players to compete with one another and test their skill in a competitive environment. As a result, Blizzard had to ensure that Diablo II could not be hacked or modified by players to give them an unfair advantage in game play. To this end, they created a client/server model, where all of the important information or valuable assets, such as player inventory, gold, skills, and player levels would be stored on Blizzard's servers, while things like results of battles, animations, and art would be run on the player's machine. Doing so allowed Blizzard to maintain a high degree of security over crucial pieces of information needed to maintain fairness in game play. In earlier online games, cheating had been rampant, mainly as a result of players being able to hack key files on their own machines to provide their characters with unlimited resources or improved weapons. One of the main selling points of Blizzard's battle.net system was that it ensured a level of fair play and balance for the player base. This would also be the central friction point that would animate much of the discourse about race and ethnicity in and around the game.

In order to do this, Blizzard assigned each player in battle.net to a "realm," a closed server which stored and secured the data for the player. Players could interact and trade with other players on their realm, but not with players on other realms. Players could choose to create characters on any of the four realms, "U.S. West," "U.S. East," "Asia" and "Europe." The realms were named geographically to accommodate player interaction based on time zone, so that each realm would have "heavy" and "light" times based on when the majority of players would log in. Heavy times on U.S. East would be 4:00 p.m. to 11 p.m. EST, while U.S. West would typically have a similar load three hours earlier.

Like most online multi-player games, Diablo II created its own economic system. Players traded valuable items, exchanged gold and loot, and players even developed a system of currency built around a game item called the "Stone of Jordan." The "Realm" system that Blizzard used kept each system closed and independent, meaning that players could not trade between realms or play with or against players on other servers, meaning that if your player found an exceptionally rare item on U.S. East and wanted to trade it (say for gold or for several other items that might benefit your character more) you could only trade with other players on U.S. East. This system was designed to make the economic system of Diablo II, more secure and less prone to cheating, item or gold duplication, and hacks. It also provided a guarantee, that virtual items would have value, setting up the possibility of a genuine economic system which could be maintained outside the game world itself in venues such as Ebay (Castronova, 2005, Dibbell, 2006).

Early on in the game, U.S. West gained a reputation as being "more serious" among players, meaning that many of the most devoted players (and therefore those likely to have the best items for trade) were playing there. Because items couldn't be traded between servers, the economies of U.S. East and U.S. West grew at different rates and created different markets based on scarcity. Over time, this created an imbalance that greatly favored US West. With more players, U.S. West saw an increase in both the diversity and number of items available, increasing the chance to acquire needed items, while simultaneously driving prices down.

As the game grew in international popularity, players from all over the world began to converge on the US West server often times overloading it and causing gameplay to slowdown or "lag." At times, when lag becomes excessive, it can interfere with game play, cause players' avatars to die and even result in the events causing players to lose items or equipment. The problem became particularly acute when international gamers discovered a large US market for Diablo II game items and began to farm them and sell them to (predominantly) US players.

Diablo II Goes Global

The problems of lag and trade imbalances became particularly acute when the game was released a year later in Korea. By 2003, Diablo II was one of the top selling titles in Korea, producing a massive influx of new players to the game. Unlike a significant portion North America's player's styles, the culture of gaming in Korea was highly competitive, resulting in an immediate source of tension. Moreover, Korean players were easily identified as "non-American" based upon language differences, which included several common, easily identifiable Korean gaming phrases, such as "Huk," which roughly translated to "Go." and more importantly by their non-responsiveness to questions or comments in English.

One significant barrier to international communication between these players was grounded in the difference in interfaces. The Korean version of Diablo II was different from the English version both in the way that it communicated information and the ways that it facilitated gameplay among Koreans.

The difference in language constituted a unique barrier to communication for Korean players. Localizations into other (particularly European) languages allowed for a shared vocabulary of game items (the name "Stone of Jordan" is recognizable in French, German, or English for example, but not in Korean). Those differences provided not only a language barrier, but also an interface barrier, in many cases making basic cross cultural communication not just difficult, but impossible. Korean players were able to create and sustain their own communities, however, around systems of "farming," creating games to repetitively farm the best areas for rare items that could be sold on third party sites such as Ebay. Many of these players were able to generate significant income based on the demand from US players for rare or high quality items in the game. Unlike US gamers, who routinely traded items to each other, Korean gamers were selling items for profit and real dollars in what emerged as a black market (selling items was against the terms of service for the game).

Within a few weeks of its release, 300,000 copies of Diablo II were sold in Korea, making it far and away Blizzard's most profitable overseas market. Because of the competitive nature of Korean game culture, most Korean players opted not to play on the sparsely populated Asia realm, but instead chose to create characters on U.S. West which had a large and established player base and trade network as well as the best market for selling and trading items.

One of the primary motives for Koreans to play on U.S. servers was economic. The economy of Diablo II found its most dramatic expression in Ebay, where items were frequently bought and sold, even though the game's End User License Agreement strictly prohibited such acts. Items which were "hot" on the realms could be bought and sold for real money on Ebay and the process became so lucrative, with items often selling for hundreds of dollars, that a cadre of "professional" Diablo II game players emerged who spent their days playing the game and item hunting, posting their items each day for auction. At the height of the game's popularity, a professional game player could make $200 to $300 a day. A significant portion of those professional gamers were Koreans, who recognized and exploited a significant American market for these virtual goods. While some American players did engage in "black market" selling of virtual goods, the primary market which emerged was one in which Korean sellers sold items to U.S. buyers.

Accordingly, to be a successful trader (Korean or Western), one needed to have access to the realms where Americans played. As the process of trading became more and more lucrative, the number of players on the U.S. West realm increased dramatically, the demand driven not by a Korean desire to flood the American servers, but by the American demand for "black market" items sold on Ebay. While a significant number of players did buy goods from

Ebay, the majority did not, setting up a complicated dynamic which was at odds with the basic design of Blizzard's system. Most players saw the "closed realm" system as a way to eliminate outside intervention in game play and many considered the purchasing of virtual goods to be cheating. Regardless, the demand for these good remained high, providing ample motivation for Korean farmers to continue to play on the US servers,creating a clear division of "turf" which would soon erupt into open conflict.

Geographies of Race and the Invention of "The Korean Problem"

As the popularity of the game began to put stress on the servers, often slowing them to a crawl, a significant portion of the player base began to racialize the problem, suggesting that it was Korean players (with whom they had difficulty communicating with in the game world due to language barriers) who were causing the slowdown. While the sheer volume of players did affect the speed of the game, there was nothing particular about Korean players which increased the latency of the network. The game's popularity had simply overwhelmed the hardware and the networks ability to manage simultaneous connections.

U.S. game players began to think of the fictional game world as a nation-space, with an accompanying sense of entitlement to the US West server domain. In the context of US culture, many of the youth playing in Diablo II would have relatively little or no direct contact with foreign national cultures. Accordingly, notions of nation and nation space, with little understanding of trade and international exchange, were codified and discussed by these groups based on the physical location of resources (such as servers and Blizzard itself) and bodies (the physical locations from where the players connected). In that sense, the physicality and geography of nation represented not only a claim to ownership, but also a possessive investment in the resources of the game as well.

In the context of the virtual world, geography becomes a marker with which youth can tie the physical body to a virtual space, especially when the physical traits of the population are markedly different, such as is the case with Asia. In that sense, Koreans provided a "target" for online racism because both their physical location and their physicality served as markers of difference. Players who were openly racist towards Korean players would have had difficulty justifying their opposition if either the markers of location or racial difference had been removed. For example, an American player connecting from Korea would be perceived as sharing a possessive investment in the virtual nation space based on racial affinity, which a Korean, connecting from the US, would be seen as having a possessive claim based on geography. It was only the combination of both location and race that challenged issues of white, American possessive investment in the game.

In Anderson's terms, the kind of nationalism that is produced by this combination of factors is typical of all notions of imagined communities, perhaps even more so when dealing with Asia. As Edward Said has illustrated, the desire to define the Other in terms of eastern and western epistemological frames is "an elaboration not only if a basic geographical distinction, but also of a whole set of interests, which it not only creates but maintains."[9] Said is describing, in powerful ways, the desire not only to define, but also to control the space of racial and geographical difference, the essence of what Lipsitz defines as a "possessive investment in whiteness." One of the most fundamental ways of defining Korean players was by demonstrating their Otherness and then framing their interests as alien as a result. Doing so allowed US players to make claims about entitlement based solely on questions of geography, allowing them to both create and mask other interests which defined the relations between and among players of different races, ethnicities, and nationalities.

Given the enormous potential for intercultural exchange signified by the game's early international reach, that possessive investment signifies something important about US West players' sense of national entitlement as an understanding of both game based and commerce based relations. U.S. Players began a campaign against Korean players—both inside the gamespace and outside on websites and forums. They used tropes of national borders and boundaries, and framed Korean players as "illegal immigrants" and "invaders." Players began joining games with Korean players with the sole intention of disrupting game play and literally chasing them off of the servers. Some players adopted racist or anti-Korean names. At one point a bug was discovered that allowed players to send a string of characters to the screen which would crash the Korean version of the game (a simple line of periods) and eject any Korean player from the game, forcing them to restart the game. As a result, they would be removed from the original game in which they were playing and be put into a queue to join other games. During the height of the conflict, it was common to see players enter a game and send the string to the screen to clear the game of any Korean players. This simple act illustrates the desire to re-colonize the space by removing Koreans and stake a claim to the available resources at any given time. Perhaps most important, while the impetus for such interaction was at base economic (Koreans playing on US West to sell items), the conflict was not. US players were in no way competing with Koreans in the marketplace.

Within the rhetoric of Korean player killing groups, there are five recurring themes which serve as commonplaces among these groups: conventions of naming designed to depersonalize and dehumanize Korean players, espousing the Western manifest destiny over the game space, the identification of Korean player behaviors as "savage" or "barbaric," attempts to distinguish "proper" and "improper" domains for the discussion of race, and explicit disavowals of racism.

When taken together, these five themes result in the creation of space which is defined by a binary of purity and impurity where any contamination is unacceptable. The space of the Western servers is implicitly defined as a white, western space and explicitly characterized as a space of "innocence."

Players of the KPK, Inc. cast themselves in the role of enforcement, as agents undertaking a mission of policing the server's space. Accordingly, KPK, Inc. created a "Most Wanted List," where players could post the names of Korean players to be hunted for "offensive behavior." The two most frequent offenses being "Clogging up the public chat rooms" and creating "Korean-Only games."

A Case Study: KPK, Inc.

As Kye Leung observed in an analysis of the game, "Sporting names such as KKA (Korean Killers of America) and FukForeigners, they blame foreigners for slowing down the U.S. servers. These racists are usually American/Canadians and White. In the chat rooms they openly admit their racist beliefs and spam the channels with words like chink and gook and that all foreigners, especially Koreans, need to go home."[10]

Typical of this kind of group was KPK, Inc., or Korean Player Killers Incorporated, a self-described "Diablo II Community Effort."11 The group which was active from 2001-2003 blamed Koreans for server instability, excessively long wait times to join games, international video piracy, creating a sense of "excessive paranoia," and filling chat rooms with "nonsense and numbers." Korean players, they argued, sought to disrupt their enjoyment of the game. As they describe the problem: "It is also all too common for a normal, peaceful, public chatroom to be instantly filled with meaningless dribble by Koreans who desire only to piss off the Western realm users."

The KPK website, which was used by a group of players to both organize Korean player killing sessions and, more importantly, to justify their actions, provides a wealth of insight into how (predominately) US players constructed Koreans in the cultural imagination. In Benedict Anderson's terms, they formed not only their own "imagined community," but also brought with it the inherent impulses of nationalism that have "made it possible, over the past two centuries, for so many millions of people, not so much to kill, as willingly to die for such limited imaginings" (p. 7).

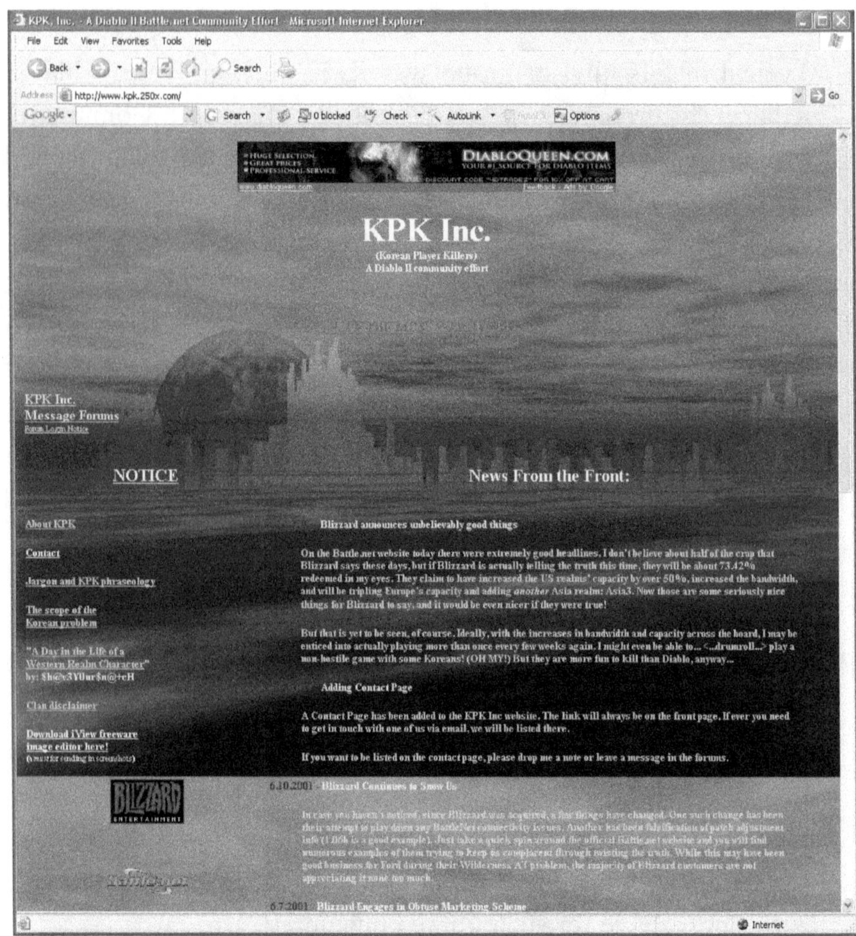

KPK, Inc. main website

Both of these impulses are alive and well within KPK, Inc. There is not only the recognition that they are "Player Killers," but also that they are also putting themselves at risk (the Korean players are as likely, or even more likely, to kill them than vice versa). The fact that the possibility of sacrifice goes unremarked on the website as a whole is interesting, but not surprising. To portray the act of killing Koreans as risky or admitting to the possibility of failure would serve to undercut KPK Inc.'s assertion of mastery over the server space.

The KPK Lexicon or You Are What You Eat

Perhaps the most direct and immediate sense of the KPK's imagined sense of race comes from the vocabulary (in their terms "special words and jargon which may or may not be confusing to the uninitiated") they develop to frame their relationships to the server space, other western players and Korean players. Perhaps most revealing is the way they code themselves. For KPK members, the term they use to reference themselves is "Fed" or "Agent." Those terms themselves suggest a notion of authority grounded in the state making them "Federal Agents," working on behalf of a broader imagined national interest, policing the borders of cyberspace and nation spaces—a legitimating discourse in their imaginations.

Interestingly, however, one's agency as a KPK Fed is not grounded in authority, but instead in a willingness to act. The very definition of a Fed, according to KPK, conflates the act of killing Koreans with a sense of legitimate authority. They write, a Fed or Agent is "[a]nyone acting on behalf of KPK Inc. in the interest of ridding the Diablo II Western Battle.net Realms of the FoK and other highly obnoxious players or groups. A Fed can be anybody: you, me, a few clan members who are bored and want to go Korean hunting (from any clan, KPK is not a clan), or anybody else who wants to take part in KPK." The explicit reference to KPK not being a clan is intended to distinguish it from groups which have regularly planned activities and formal membership requirements. In contrast to a clan, one becomes a member of KPK Inc. simply by killing Korean players.

Of particular note is the reversal of the traditional relationship between authority and action. Agents, particularly working on behalf of governments, are traditionally given sanction to act by the state in ways which would be illegal in other contexts, such as detaining or killing people. In the case of the KPK, acting in such ways (in the service of what they see as the greater good) is performative, in the sense that the act of killing Koreans conveys, automatically, the authority to do so. As a result, "anybody who wants to take part" (presumably as long as they are not Korean themselves) is already a member.

The labeling of Koreans themselves is much more straightforward. According to the "Jargon and Phraseology," Korean players can be identified with the following designations:

> FoK (Forces of Kimchi). Kimchi is a special Korean food. This is used to refer to one or many Koreans (as in: "that FoK" or "the FoK came after me again!").
>
> WOA (Way Out of Area) This term should be obvious. The Koreans who clog up the Battle.net realms outside of the Asia net are this.
>
> CEA (Cabbage Eaters' Anonymous) Similar in usage to "FoK".
>
> I-Plz. This term refers to one single Korean player. It is short for "Item plz!!!!!!" which is something any B.net player is all too familiar with seeing. It seems the only English that they have learned is "Item Plz!!!", "English, Japan FUCK FUCK!", and "Give me money!!!". I haven't a clue

> where "gogo" comes from, but it sure is annoying when the whole screen fills up with it.
>
> Gogo. Another singular term for a Korean player. Again, anybody who has played on Battle.net must be familiar with this, as they write it everywhere possible for totally inexplicable reasons.

The use of food in racial slurs is quite common (even whiteness is signified by “White bread” and “crackers”), so it is hardly surprising to find food as a marker for race. Kimchi, especially, is a designation which metonymically marks the nation of Korea as a kind of national cuisine. In that sense it is both localized in a nation-space and fundamentally excluded from the “proper” space of U.S. Western Culture. In that sense, consumption of a particular food is read as both alien and obsessive. Koreans do not merely eat Kimchi, they are defined by the practice of eating it such that it is read as fuel (Forces of Kimchi) as well as an obsession that requires treatment (Cabbage Eaters Anonymous). It is an effort to both depersonalize and dehumanize the players as alien others with strange (and possibly dangerous) obsessions with things we don’t understand, while at the same time, disempowering them through characterizations which have no power within the context of the game.

The corollary to game behavior functions in similar ways, demarcating behaviors which are to be read as pathological: being “way out of area” and hence occupying spaces which are inappropriate, making no effort to learn the language of the space they occupy, and being either crazy and/or inscrutable, obsessively spouting “Gogo” for “totally inexplicable reasons.”

As a result, Korean players are never seen as players at all, but instead as noise in the system that cannot be understood rationally, cannot be reasoned or communicated with, and who have no proper place within the game world. In short, it is an effort to call forth the Derridian principle, "tout autre est tout autre," that every bit of them is every bit other.[12]

Such arguments bear a striking similarly to the discourse on race and racism that permeate U.S. culture with respect to issues of illegal immigration, where people are reduced to food ("beaners") or practices ("wetbacks") and are seen as criminal, irrational, or other to such a degree that the only understanding of them that is possible is their identification as wholly and completely other.

Manifest Destiny

The primary arguments that KPK, Inc makes to justify player killing reflects a belief in a kind of manifest destiny, with Blizzard, the developers of Diablo II, cast in the role of the divine creator. For KPK, Inc., the question of realm ownership, rights and privileges is clear cut. Their use of racist terms, they claim, is not designed to offend, but instead to make clear to people who does, and more importantly, does not belong in realm server space. As they submit, "These phrases, while somewhat derogatory in nature, are intended only as a slang means of expression that can be generally accepted by everyone in an effort not to offend or incite, and are not aimed at Korea or Koreans as a whole. They are intended solely to label the Korean players who are WOA (Way Out of Area) and are annoying to the point of spoiling the game for countless innocent Western players whom the US West, US East, and Europe Realms were designed explicitly for."

It is clear to KPK, Inc. that they have a certain entitlement to server space and that their language, while "somewhat" derogatory in nature is not meant to offend, but to label. The primary objection is not about player behavior, but instead, it is about presence. It is the very fact that they are on the server ("Way Out of Area") that KPK, Inc. finds objectionable. Thus KPK, Inc develops a strategy for marking those who are "spoiling the game" for "innocent Western players."

What gets conflated in this notion, however, is the idea that presence itself constitutes "bad behavior." If you are recognizably Korean and you have a presence on one of these servers, regardless of what you do, you are ruining the experience for others. Moreover, for KPK, Inc. the issue is a foundational one: being Korean violates the design principle. As a result, what is to be banned is not any particular act, but rather the performance of an identity.
The question at stake is one of manifest destiny, not only who owns the space, but who owns the rights to the space. KPK, Inc. embodies a kind of American exceptionalism that they use to justify their extremism. As the voice that can be "generally accepted by everyone," they speak from a position of generality and authority, echoing earlier incarnations of manifest destiny and engaging in external aggressiveness characteristic of "race patriotism."[13]

Ultimately, the goal of KPK, Inc.'s vitriolic rhetoric is to establish a dichotomy between Western and Korean players which would enable what Reginald Horsman describes as two of the most powerful connectors between race and manifest destiny: making the victim responsible for their own destruction and creating an overarching rationalization that makes the group, rather than the individual, responsible for racist and potentially genocidal behavior.[14]

Savage Behavior

A significant portion of the KPK, Inc. website is used to label, mark, and describe the behavior of Korean players, in an effort to categorize them as other. In doing so, KPK Inc. replicates the cultural logic of imperialism. Aware that the initial perception may cause concern, KPK Inc. wrote several documents defending their rationale for Korean player killing. "For anyone not familiar with playing Diablo II on Battle.net in the Western Realms," the document begins, "a website dedicated to PKing Korean players on the grounds that they are out of their area or are simply an annoyance may seem a bit extreme, or even plain mean."

What follows is a set of distinctions designed to distinguish the proper from the improper and that distinction relies exclusively, in the KPK Inc. rhetoric, on notions and understandings of place. While physical markers of race and ethnicity are absent in the virtual world, cultural and geographical markers abound. Not surprisingly, many 20th and 21st century American sources of racism have been about place as well. From Rosa Parks refusing to sit in the back of the bus, to disallowing illegal immigrants entry to hospitals and schools in California's Pop 187, racism in America has had as much to do with the physical location of people's bodies, as it has to do with other factors.

The primary argument that KPK proffers in defense of killing is that Koreans are, quite simply, out of their place. "The Koreans who are playing outside of Asia" they argue, "are by no means nice, reasonable or in any way fun to play with (with a very few notable exceptions)." Their behaviors, according to KPK are one dimensional and well known: "As all Diablo II players know all too well, they are famous for trying to ransom items that you accidentally drop, filling up the screen with repeated calls for "Item plz!" while you are heavily engaged in battle, not allowing you to talk to NPCs in a high-lag game by following you around and trying to trade incessantly (only to have nothing to trade nor anything to say once the trading screen comes up accept "Item plz!" or "Gold!!!")."

The point of such descriptions is not to argue that the behavior itself is problematic, but that it reflects a savage temperament, which further marks Korean players as out of place. They lack propriety and, as a result, should be excluded from having a sense of place. It is not merely the interaction that is a problem for KPK Inc. They are also disturbed by Korean players' isolation. "Korean players are also famous for starting private, 1 player games for a few minutes, leaving, and starting a new game. Why would that hurt anybody? Well, all the legitimate indigenous players on the Western Realms are having to wait up to 20 minutes in a game-creation que[sic] just to play! It also contributes to server splits." Of course, one cannot help but notice the deployment of the term indigenous to describe the players of the U.S. West realm. It is a term which connotes both historical claim and a geographic/spatial connection to the space in question. Most importantly, as a location of privilege, the notion of being a native, as the word indigenous suggests, conveys a birth right, something to which US players have no more claim than any other player who is "born" on that server.

Finally, KPK, Inc. offers what they see as the ultimate solution, which is the virtual extermination of Korean players. By repeatedly killing them and thereby interfering with their game play (in essence doing precisely what they accuse Koreans of doing to them), KPK, Inc wants to drive them back to what they see as their proper place, the Asian realms. "All we want to do," the site explains, "is

PK [player kill] them enough times in a row that they are convinced to stop playing outside of Asia."

The right, they claim is geographically based, "Within the Asia realm, we Westerners are the intruders and must abide by their rules. When they are in our realms, they must abide by our guidelines and rules of decency" and is dependent on players to enforce. While Blizzard could have easily set an ip address filter to distribute players geographically to regional servers, they allowed players to connect to any realm. Doing so facilitated players connecting with friends or other players in different time zones or in other parts of the world.

One of the primary motives for Koreans to play on U.S. servers was economic. The economy of Diablo II found its most dramatic expression in Ebay, where items were frequently bought and sold, even though the game's End User License Agreement strictly prohibited such acts. Items which were "hot" on the realms could be bought and sold for real money on Ebay and the process became so lucrative, with items often selling for hundreds of dollars, that a cadre of "professional" Diablo II game players emerged who spent their days playing the game and item hunting, posting their items each day for auction. At the height of the game's popularity, a professional game player could make $200 to $300 a day. A significant portion of those professional gamers were Koreans, who recognized and exploited a significant American market for these virtual goods.

Accordingly, to be a successful trader, one needed to have access to the realms where Americans played. As the process of trading became more and more lucrative, increasing numbers of Korean players began playing on US servers. The conflict between US and

Korean players was framed differently for Blizzard than it was for KPK, Inc. and other players who shared their beliefs.

While players could have easily petitioned Blizzard to shut down accounts which were violating the terms of service for the game by selling items on Ebay, they chose a different tactic, preferring instead to take matters into their own hands. As the website explains it: "Blizzard has, of course, anticipated all of these problems and arguments about realms, jurisdiction, etc. and has decided to stay out of the arguments and disputes between and amongst players. Therefore, it is fully fair and decent within the rules of the game and the game world of Diablo II for us to band together to fend off the irritating, offending players."

Because Blizzard had opted to stay out of the conflict, players not only felt the need to take matters into their own hands, they felt perfectly justified in doing so within the framework of the game itself. Blizzard, they felt, had provided them with the necessary tools to defend their borders as they had constructed them in their own imaginations. By implication, if they were simply following the "rules of the game" and being "fair and decent," it would be impossible to accuse them of being racist, without implicating both Blizzard and the mechanics of the game itself as racist as well. Doing so marks the first step in the strategy of disavowal which is designed both to avoid charges of racism and insure the maintenance of a possessive investment in white privilege.

Race, Place, and Disavowal

KPK, Inc., is clear aware that their words, actions, and rhetoric will be read as racist. Accordingly, they go to great lengths to explain what racism is and why their discourse should not be read as racist.

Central to their discussion is the formulation of racism as an "all or nothing" proposition. Racism, they contend, has a certain absolutist quality, which must be both "open" and "100% derogatory." Such conditionality provides a space for disavowal through possessive investment. The definition of racism is no longer defined by either the content of the discourse or the effect that such language or actions may have on others. Instead, racism is defined, or rather negated and erased by uses of counter-example. To prove one is not racist, all one must do, according to KPK, Inc. is provide a counter-example to racism.

As their website warns visitors:

> "As was stated on the NOTICE page, this site is not racist, nor do we condone open racism. Your personal views are your own, of course, and nobody can stop you from feeling or thinking whatever way that you do. However, on this site, we do not and will not condone the use of openly racial slurs which are popularly recognized as being 100% derogatory. Please refrain from the use of words such as: Gook, Dog-Eater, Frog-Head, and Slant-Eyes. They imply much more than annoying B.net behavior and also indicate a distaste for all Asians, not just Koreans.

Let's not miss the point here, we are only concerned with Diablo II on Battle.net, not world issues or race. Also, please remember that it is within KPK Inc. policy to place a non-Korean on the Most Wanted List, and really, this site is all about PKing the most annoying bastards on B.net who try to spoil the fun for the rest of us. It just so happens to be that 99% of them are Koreans, thus KPK."
The "notice" they provide reads like a primer in how to avoid accusations of racism, while outlining permissible permutations of racist discourse. In fact, the only prohibition explicitly stated is

against open racism, adding that "of course, nobody can stop you from feeling or thinking whatever way you want to do." One many not use "openly racial slurs which are popularly recognized as being 100% derogatory," but one must instead target racial slurs more specifically to Koreans.

The "Notice" is designed to reinforce the two central arguments that KPK Inc has put forward in defense of racism: the specificity of Koreans who are "out of place" and the notion of "savage behavior."

To return to Lipsitz's notion of possessive investment, it is important to note how preemption and disavowal function in unison to redefine notions of social relations as they are constructed around race. Here KPK, Inc. offers a series of disavowals which reflect three distinct strategies: Disavowal by example, disavowal by circumstance, and disavowal by denial.

The rhetorical strategy of argument by example takes what the author believes is compelling evidence which logically precludes them from being racist. Often this kind of disavowal will center around acquaintance or friendship, their connection to a person of color offering "proof" that they are not racist, after all, how could a true racist count a black, latino, or asian person as a friend?

The head of KPK, Inc. extends the logic one step farther in arguing that "KPK Inc. is in no way racist, nor do we support racism. [the site owner] has absolutely no problem with Asians in general and is, in fact, married to one." While the connection between friendship and marriage on one hand and cultural attitudes towards race on the other are tenuous at best and non sequiturs at the worst.

The second argument suggests a broader social or cultural context which provides a rational for racist behavior. It is a "racism happens" argument, the result of conflict, emotion, and lack of effective borders: "As people from different cultures, backgrounds, social castes, races and religions have come into contact with each other throughout history, there has always been an element of distrust and conflict. It is only natural that this same situation would exist on the Internet where there are no effective borders, and especially in the online gaming world where the very games we play against each other direct us to compete and fight some way. Generally speaking, people of like backgrounds tend to stick together, and in these games the situation is no different. Unfortunately, the emotions present in the game world sometimes surpass those that were intended by the game designers, and it is possible for these heightened emotions to take the form of open racism."

Finally, the site owner's disavowal is completed through denial (in the double sense of the word): "Here at KPK Inc. we absolutely do not support, condone, or acquiesce to any strictly racist views. Of course, each and every individual is welcome to his view of the world, but expressions of pure racism are not permissible and are, in the end, truly counterproductive."[15]

Conclusions

The case of Diablo II is, in some respects, a familiar performance of national and racial adherence, employing classic tropes and forms and mapped onto an unusual but still rather literal representation of space and territory. And yet the strangeness of the circumstances brings out the episode's formal character—the need for an identity principle that can define the ingroup ("people of like backgrounds tend to stick together, and in these games the situation is no different") despite the manifest difficulty of knowing who one's compatriots are. Or the overproduction of racialized injury on an extremely thin basis of social contact. On the one hand, this formal character is testimony to the portability of the race/nation discourse: it should come as no surprise that it can be asserted in virtual spaces, or that investments in virtual lives should give rise to real senses of injury. On the other hand, the formal character is suggestive of the way these events come to resemble a game within the game—with rules, a narrative, forms of action, and arguably less requirement that the racist overflow be integrated into a larger racist worldview, with outside-the-game implications. Such games within the game develop through the processes of player-constructed meaning and action that always surround the gamespace, no matter how scripted. If this margin is responsible for a great outpouring of new roles and forms of cultural engagement—of playful interventions—it should come as no surprise that it also produces playful hatreds.

As massively multiplayer online games advance, so have the arguments about notions of nation, space, and race. Games such as Lineage and Lineage II have produced heated conflict over race, nationality and playstyle (Steinkuehler, 2006) and most recently World of Warcraft has responded to player complaints of Chinese players "gold farming" by banning hundreds of thousands of accounts from the game. Many of the arguments proffered in support of killing Korean players in Diablo II have found their way into the gamer lexicon and have been applied to Chinese players working for companies which farm items and gold in most of the larger multiplayer games.

Throughout this essay, my chief claim has been that we should look to issues of power, privilege and investment as markers of race in cyberspace and that often these issues are manifested in strategies of denial and disavowal, rather than explicit racism. All too often, our temptation is to look for the physical markers of race and ethnicity, markers which are easily erased or submerged within the non-physical space of virtual worlds. That does not mean that race, ethnicity, or racism have disappeared, but rather that each has been transformed and in some cases radically altered within the context of the net. As we examine these issues, we need to pay careful attention not only to the representations of race and ethnicity as they appear on the surface, but also to the emergent cultures that spawn around these images and representations. Doing so can help us understand not only what is being learned in the context of new media, but how issues of race and ethnicity are being woven into the fabric of culture through the language and practices that people use in their responses to difference and change.

References

Anderson, B. (1991). Imagined Communities: Reflections on the Origin and Spread of Nationalism. New York: Verso.

Castronova, Edward. (2005). Synthetic Worlds: The Business and Culture of Online Games, Chicago: Chicago UP.

Dibbell, Julian. (2006). Play Money: Or, How I Quit My Day Job and Made Millions Trading Virtual Loot, New York: Basic Books.

Derrida, J (1996) The Gift of Death, Chicago UP. p. 68

Everett, A. (2005). "Serious Play: Playing with Race in Contemporary Gaming Culture." Handbook of Computer Game Studies. Eds. Joost Raessens and Jeffrey Goldstein. Cambridge, Massachusetts and London, England: MIT Press. 311-326..

Horsman, R. (1981) Race and Manifest Destiny: Origins of American Racial Anglo-Saxonism, Cambridge: Harvard UP, 1981.

Kolko, B. (2000). Erasing @race: Going White in the (Inter)face. In Race in Cyberspace, NY: Routledge, 2000.

Lipsitz, G. (2006) The Possessive Investment in Whiteness: How White People Profit from Identity Politics, Temple UP.

Leonard, D. (2006). "Not a Hater, Just Keepin' It Real: The Importance of Race- and Gender-Based Game Studies," Games & Culture, 1(1): 83-88.

Said, E. (1973). Orientalism. New York: Vintage.

Seiter, E. (2005). The Internet Playground: Children's Access, Entertainment, and Mis-Education, New York: Peter Lang.

Steinkuehler, C. (2006). The mangle of play. Games & Culture, 1(3), 1-14.

Takaki, R (1990). Iron Cages: Race ad Culture in 19th Century America, New York: Oxford UP ,1990, p. 269.

Taylor, TL (2006). Play Between Worlds: Exploring Online Game Culture, Cambridge: MIT Press.

Thomas, D. and Brown, J.S. (forthcoming, 2007). "The Play of Imagination," Games & Culture, 2(1).

Notes

[1] Beth Kolko, Erasing @race: Going White in the (Inter)face. In *Race in Cyberspace*, NY: Routledge, 2000.
[2] Lipsitz, p, viii.
[3] See for example, Ellen Seiter's *The Internet Playground: Children's Access, Entertainment, and Mis-Education.* New York, Peter Lang, 2005.
[4] For a discussion of the need for this research see, Leonard, 2006 and Everett, 2005.
[5] For an extensive reading of one such game culture, see TL Taylor's *Play Between Worlds.*
[6] See for example recent work done on gaming in international contexts.
[7] http://www.bluesnews.com/cgi-bin/articles.pl?show=44, sales figures are quoted from PC Data, the industry standard for monitoring and tracking PC games sales figures.
[8] http://www.blizzard.com/diablo2/ Last accessed August 2, 2006
[9] Said, 1978, p. 12.

[10] *Kye Leung, "Asians and Asian Americans face racism in online game," Azine, last accessed, July 18, 2006.*
[11] http://www.kpk.250x.com. All references to the KPK are from this website, last accessed April 28, 2006.
[12] Jacques Derrida, *The Gift of Death*, p. 68
[13] .Ronald Takaki, *Iron Cages: Race and Culture in 19th Century* America, New York: Oxford UP ,1990, p. 269.
[14] Reginald Horsman, Race and Manifest Destiny: Origins of American Racial Anglo-Saxonism,Cambridge: Harvard UP, 1981.
[15] KPK, "Racism and Multiplayer NetGames," ibid.

Chapter Four: Viral Style: Technology, Subculture, and the Politics of Infection

The phenomenon of the computer virus is something that nearly every computer user is familiar with, either via direct experience, through media and software advertising, or from stories that permeate the popular imagination. While each of these aspects describes an element of computer viruses in contemporary culture, none adequately explains the historical, social, or political contexts of computer viruses themselves, of the programmers who create viruses, or of the multi-million dollar industry that has been created to protect computer users from them. Both the underground production and dissemination of viruses and the industry that has emerged to counter that activity are products of the computer networks that host them and the moment of definition when it became possible to speak of a "computer virus." That moment originates in the discourses of science fiction in the 1960s and early 1970s, but became realized in the 1970s and 1980s in the world of computer science with the definition, formalization, and study of computer viruses.

The origins of both the underground culture of virus writers and the discourse of computer viruses in the popular imagination can be traced back to the history of computing, the history of science fiction, and the moment of articulation regarding the "invention" of the computer virus. In this essay, I theorize the culture of virus programmers and the cultural responses to viruses that they produce by examining how the idea of the computer virus evolved in the multiple discourses of computer science, medicine, and law. What such an examination reveals is that the social and political motivations that underlie the culture of virus writing is directly related to the transformation of the discourse of computer viruses from a discourse of technology into a discourse of biology initially, and later into a discourse of hygiene. The transformation of viruses from technological entities to biological ones gave rise to system of medical perception in relation to computers and computer culture. Ultimately, the trope of hygiene was deployed within computer culture itself as a means to understand viruses and their effects as well as to enact a system of normative, and later legal, system of social control. In response to those transformations, virus production can be seen as an act of resistance in an era when computer culture is becoming increasingly commodified and when the discourse of computer programming is undergoing a process of normalization and exclusion.

From Technology to Biology

The technological and cultural histories of viruses are squarely situated as an irreducible component of technological functioning. Viruses are neither alien to or separate from computers, but, instead, an essential part of their operation, indistinguishable from other types of programming apart from their effects. In that sense, the computer virus is *fundamental* to the computer as it is unable to be separated from the basic nature of the computer's functioning. The moment in the early 1980s when the idea of the computer virus was abstracted and divorced from the idea of the computer marks a critical junction in the discourse of computers and computing. That divorce represented the construction of a discourse of disease, infection, and pathology that has further necessitated the invention of a kind of "medical perception."

The transformation of technology into biology carries with it a set of implications about the social nature of computing as well. As computers became networked, there arose a need to find a new vocabulary to describe not only the technology, but the interactions of various technologies. Metaphors from the world of biology provided a ready made set of terms that could provide perspective on the rapid changes that were resulting from new technologies. As Len Adleman, one of the first computer scientists to create and name a computer virus, described the environment, "Computer viruses are an inevitable part of computer networks, because computer networks are 'substrates' for evolution. If things can evolve they will evolve. We're back to biology. Computer viruses are things that 'belong' on computer networks. They are as natural for that environment as biological viruses are for our physical environment."[1] This fit between networks and ecosystems provided a connection that suggested a progression based on the concept of evolution. Viruses, like biological organisms, would flourish, change, grow and develop within the technological ecosystem just as they would in the biological world. The metaphor of evolution would provide the basis for the early research into computer viruses, work done by Adleman and his student Fred Cohen in the early 1980s.

While the discourse of evolution provided a theoretical structure for understanding viruses, it failed to translate into either the larger computer community or the popular imagination. Instead, Adleman, the discourse that has evolved around the computer viruses treats them as anything but natural. Instead, viruses are seen as contamination, as an infection in an otherwise pure system or environment. This shift, which marks a change in the discourse from evolution to contamination, is also the point at which the discourse of computer viruses becomes public.

The transformation which occurred in late 1980s produced a proliferation of discourse around the term "computer virus," whereby the terms meaning was both reconfigured and contextualized in terms of its public nature. As Foucault marks this kind of change, it is the moment when a "mutation in discourse" takes place. Prior to that moment, Foucault argues, "things and words have not yet been separated," the point where "at the most fundamental level of language–

seeing and saying are still one."[2] With the metaphor of evolution, viruses are discussed in the same terms as other programs and understood in the language if mathematics (particularly "set theory" as described below). The primary question is one of definition, which asks how *within the discourse of technology* can we differentiate viruses from other types of computer programs. Indeed, as Adleman describes viruses themselves, "it is possible to think of the virus as a map from programs to ('infected') programs."[3] In that sense, viruses mark the relationships between programs, or more precisely between programs and "infected" programs. Such an assumption is a clear indication of the manner in which it is not only possible, but even preferable, to think of viruses in terms of the technology which they infect. Infection, then, is not something alien to computer programs, but it is what marks a particular kind of intersection of two different programs. It is also the context in which we can begin to understand computer viruses not in terms of definitions, but in terms of their nature as entities on a network.

The Perfect Rhizome

The notion of mapping which Adleman describes is not altogether unfamiliar theoretical territory. What Deleuze and Guattari describe as a rhizome offers an ideal theoretical construct for the mapping on the virus both in technological and social space. The rhizome is, in this context, understood as a articulation of multiplicity without unity, a construction which is defined by a multiplicity of connections, rather than through a singularity, definition, or deep structure. In short, the rhizome is a network of interrelated connections and lines of flight, it is a mapping of movements which perform acts of continual deterritorialization and reterritorilazation. Deleuze and Guattari use the image of the orchid and the wasp to illustrate the principle: "The orchid deterritorializes by forming an image, a tracing of a wasp; but the wasp reterritorializes on that image. The wasp is nevertheless deterritorialized, becoming part of the orchid's reproductive apparatus. But it reterritorializes the orchid by transporting its pollen. Wasp and orchid, as heterogeneous elements, form a rhizome."[4] The movement is identically expressed in the movement between computer and virus, whereby the virus both reterritorializes the space of the computer and is itself deterritorialized as it moves and is moved throughout the network. The problem of defining the computer virus stems from precisely this dual nature, both its deterritorializing and reterritorializing force.

As Deleuze and Guattari point out, understanding the rhizome is not a matter of definition in the traditional sense (that of reproduction). Instead, the rhizome is better understood as a problem of cartography. As such, they contrast two cartographic epistemologies: tracing and mapping. Tracing is an epistemology of representation and reproduction. Its "goal is to describe a de facto state, to maintain a balance of intersubjective relations. . . .It consists of tracing, on the basis of an overcoding structure or supporting axis, something that comes ready made."[5] Tracings articulate the definable space of knowledge which is, literally, rooted in structure and which can be broken down into constituent parts, labeled, categorized and defined. By way of contrast, the rhizome is "altogether different, a *map and not a tracing*."[6] Unlike the rigid epistemology of the tracing, the map is "open and connectable in all of its dimensions; it is detachable, reversible, susceptible to constant modification. It can be torn, reversed, adapted to any kind of mounting, reworked by an individual, group or social formation. . . .A map has multiple entryways, as opposed to the tracing, which always comes back 'to the same.'"[7]

Like the wasp and the orchid, understanding the relationship between the virus and the computer can be best understood not as a problem of definition, but as a question of cartography. Indeed, the nature of the rhizome as both connected and heterogeneous marks a break in the nature of signification: "any point of a rhizome can be connected to anything other, and must be," Deleuze and Guattari argue, but "not every trait in a rhizome is necessarily linked to a linguistic feature: semiotic chains of every nature are connected to very diverse modes of coding (biological, political, economic, etc.) that bring into play not only different regimes of signs but also states of things of differing status. *Collective assemblages of enunciation* function directly within *machinic assemblages*; it is not impossible to make a radical break between regimes of signs and their objects. Even when linguistics claims to confine itself to what is explicit and to make no presuppositions about language, it is still in the sphere of discourse implying particular modes of assemblage and types of social power."[8] In short, the very definition of a virus (as Adleman points out) is not that of a de facto state or object, but, instead, is an expression of movement, a map from one infected program to another.

That mapping, however, is not limited to the domain of the computer, but, like all rhizomes, the map extends into the realms of the social, cultural and political. What follows is an effort to provide a mapping of viral movement through the space of medical perception, popular cultural and the law, not as distinct and separable domains, but as cartographic spaces of deterritorialization and reterritorialization, as multiple pathways into an through the rhizomic structure of infection. That structure is an attempt to defy the various regimes of signs which seek to contain, codify and contain the notion of the virus.

In order to map the computer virus, we first need to understand the space of de and reterritorialization in which viruses exist. Specifically, the computer virus needs to be understood in terms of the basic (even essential) function of computation: repetition. Such a mapping discloses not only the technological space of the computer, but also the social space of computer culture.

Repetition as the Nature of Computing

The power of computers, as is the case with all technology, is in the ability to repeat. Technology, broadly defined as "a way to do things," is nothing more that the codification of a set of procedures or the production of an instrument which can be used repeatedly. The same is true of all computer languages, algorithms, code, and programs. Their values are not in any particular function, but in the ability to perform a function repeatedly. There is nothing a computer can do, at least in the abstract, that a human cannot. Computer simply do things much faster and more accurately. The value of the computer is in repetition, without error and at increasingly higher rates of speed.

The basic structure, inherent in all computer languages, is the idea of "the loop," a command which tells the computer to repeat a certain set of instructions either indefinitely or until a certain condition is reached. It is in the context of this looping that computers become particularly valuable to us. Once programmed, a computer can take a set of variables and perform the same operation for each variable repeatedly, measuring differences and similarities between them. The computer is indifferent to results or data; it functions in the same manner, regardless of the information that it is either given or that it produces. The power of the computer resides in its ability to change or enhance human perception. Perception of the observer is at the heart of the meaning of repetition. As Deleuze remarks, citing David Hume's dictum, "repetition changes nothing in the object repeated, but does change something in the mind which contemplates it."[9]

This sense of contemplation is the basis for what Sherry Turkle has called the culture of calculation and the culture of simulation, which have each produced aesthetics for the high-tech world of computers.[10] For Turkle, the ability to represent "reality" on the surface-space of computers marks a shift from a culture of calculation to one of simulation. In calculation the function of repetition is clearly marked and well understood, technology functions transparently, providing access to the inner workings of the machine. In a culture of simulation, where the technology and its functioning is rendered "opaque," repetition is hidden. The goal of simulation is not to provide information which is useful to better understanding reality. The goal is to represent reality with the computer. Repetition, as the abstract series of calculations without difference, is replaced by representation and simulation.

The worlds of calculation and of simulation differ in a second, more important respect. While calculation takes "repetition" as its master trope, it remains abstract, contained inside the computer and to a large extent separated from reality. Calculation allows us to reflect on the world from a different domain, it allows us to take the words and numbers off of the screen and translate them back into the world. In short, calculation as a world of pure repetition, allows us to contemplate the world as if it too was capable of pure repetition. In the world, however, things never repeat, not perfectly. In the digital world, where everything is reducible to 1s and 0s, perfect repetition is possible. The world outside of computers, however, is governed by analogy. It is a world in which things look like one another, but never repeat precisely. It is an analog world, where relationships are governed by proportion and difference, rather than the absolutes of on and off. In simulation, repetition (1s and 0s, repeating without difference) is replaced with by representation (analogy, repeating with difference).

Such a shift gives different meanings to the concepts of repetition, particularly when repetition is elevated to the next level, that of reproduction. The idea of a computer program or code reproducing itself is also a fundamental component of a computer's functioning. Copying files, code, or data are all crucial elements that make computers useful tools. Digital reproduction is, itself, defined as perfect replication, as reproduction without difference. That notion of reproduction is also impossible in the analog world.

In the culture of calculation, reproduction (even self-reproduction) is nothing more that the reproduction of bits, of 1s and 0s. It is an act of pure repetition, which marks digital technology as distinct from its analog counterparts. It is able to copy itself identically and completely without difference. The translation from the culture of calculation to the culture of simulation, transforms reproduction into a different event. In the analog wold, only biological entities are capable of reproducing themselves. As a result, copying and reproduction as simulation are understood in terms of their analog equivalent–Life. In blurring the line between the digital and the analog, repeating, the essential function of computing, is translated into biological reproduction.

The transformation from calculation to simulation provides the context in which it first became possible to speak of a "computer virus" in terms of infection, rather than evolution. It also produced the environment which would necessitate a different kind of perception for how computer programs generally and viruses specifically would be understood.

Defining Viruses: Computer Culture, Abstraction and Difference

The idea of the computer virus was in circulation in science fiction for decades before it was put into practice on actual machines. As Scott Bukatman argues, the idea is nascent in the works of William Burroughs in the 1960s and was heavily referenced in relation to the image in other works of science fiction in the 1970s and early 1980s.[11] The earliest accounts of actual virus programming date back to the early 1970s, when programmers at the Lawrence Livermore lab began experimenting with self-reproducing "bad code," which they saw as a potential threat to computers and the ARPANET (the precursor to today's Internet).[12] In his experiments, Gregory Benford, a research physicist at Lawrence Livermore labs created two programs, one called *virus* and the other called *vaccine*, and the concept was later picked up and developed into several short stories by science fiction author (one of Benford's friends), David Gerrold.[13] The moment of transition from the "stand alone" computer to the idea of the network, or the creation of the "social computer," is marked, then, as the moment where infection become possible. The idea of the network makes it possible to envision the transmission of disease through the computer and marks one of the origins of the technological transformation of computers into biological entities.

The metaphor which governed this initial description of a computer virus was still evolution. In fact, the climax of Gerrold's early stories (and a subplot which later omitted from the book) concerns the evolution of the virus program as a result of telephone line noise. The virus is mutated and the vaccine program marketed to stop it is rendered ineffective as a result of the virus being let loose on the network. It literally evolves.

Evolution and the Invention of the Computer Virus

The formal definition of computer viruses would occur in 1983, through a collaboration between Leonard Adleman and his student at the time, Fred Cohen. As Adleman points out, "the term 'computer virus' existed in science fiction well before Fred Cohen and I came along. Several authors actually used that term in science fiction prior to 1983. I don't recall ever having seen it, perhaps it was just a term whose time had come. So I did not invent the term. I just named what we now consider computer viruses 'computer viruses.'"[14] Cohen's PhD thesis, *Computer Viruses*, attempted to create a formal definition of a computer virus but met with limited success. The primary problem with Cohen's description was that he was unable to separate the idea of the computer virus from the environment in which it operated. His definition would turn out to be best expressed as a mathematical model. In short, there was no clear point of breakage between the thing and its description that would allow it to be put into a discourse separated from viruses themselves. In that sense, Cohen's work was more an abstraction than a definition. What was put into language has been generally dismissed as inadequate. As the VIRUS-L Frequently Asked Questions text describes it, "Cohen's formal definition (model) of a virus does not easily translate into 'human language,' but his own, well-known, informal definition is 'a computer virus is a computer program that can infect other computer programs by modifying them in such a way as to include a (possibly evolved) copy of itself..'"[15] That definition, along with most others which have been offered, has been soundly criticized for its inability to differentiate viruses from other more useful programs. Apart from imposing external criteria (such as effects or arbitrary limiting factors), viruses cannot be easily separated from other classes of programs. The problem is not defining what a computer virus does. Those aspects were well understood. The difficulty is in defining a computer virus in any exclusive way, creating a definition that defines a virus and excludes programs or software that are not viruses. As Eugene Spafford describes the problem, "Cohen's formal definition includes any program capable of self-reproduction. Thus, by his definition, programs such as compilers and editors would be classified as 'viruses.'"[16]

The formal definition of a computer virus, according to Adleman and Cohen, is expressed in set theory.[17] In such a formal definition, the idea is to define what, precisely, constitutes a "viral set" for a given machine and to do so in such a way that the set is also general in its definition. Set theory is primarily concerned with both the questions of how one defines the conditions under which something can be said to be an "element of" a given set as well as the scope and boundaries of what constitutes a "set" itself (e.g., can there be said to be a set with an infinite number of elements). Late 19th and early 20th century mathematicians such as Georg Cantor and Kurt Gödel and philosophers such as Bertrand Russell, Alfred Whitehead, and Gottlob Frege (as well as a number of others) had worked from the premise that mathematics is reducible to logic and that through various axioms, one can determine what is and is not logically able to be contained within a particular set.

For the abstract definition of a computer virus Adleman and Cohen set out to prove that there could be such a set, a "viral set," which contained programs which modified other programs "so as to include a (possibly 'evolved') version of itself."[18] The central theme that Cohen addressed in his doctoral research was not merely identifying viral sets, but accounting for evolution within them. "Several previous attempts at definition failed," he notes, because earlier assumptions about the singular nature of viruses made "the understanding of the evolution of viruses very difficult."[19] The difficulty in definition and the necessity of using set theory to describe viruses is a product of their evolutionary nature. In the mathematical proof, Cohen explains, the very definition of the viral set "embodies evolution" to such a degree that the transmission of the virus is essentially an act of evolution. "Evolution," Cohen writes, "may be described as the production of one element of a viral set from another element of that set."[20]

These early (and most successful) attempts at definition, defined viruses as mathematical entities, logically tied to the machines and programs which they inhabited. Their evolution was similarly tied to the mathematical and computational environment in which they were born and reproduced. The metaphor of evolution leads to another conclusion, which further illustrate the symbiotic relationship between virus programs and the networks they inhabit. In order to completely neutralize a virus, it must be able to be put in a state referred to as "absolute isolation," meaning that the spread of infection can be stopped and the infected program can be detected and removed because it is isolated from any other contact on the network. As Adleman demonstrated, however, not all viruses are "absolutely isolable,"and as a result, within the computing environment, "protection cannot be based upon deciding whether a particular program is infected or not."[21] Adleman's proof means that once a network is infected, providing that a particular type of virus has infected it, it is impossible to ever guarantee that the infection itself is completely eradicated. Put simply, one a network is infected, it can never be cleaned.

Adleman and Cohen's work both focused on the elegant expression of viruses nature in the language of logic and mathematics. In a more practical sense, the difficulty with defining viruses comes not from the mathematical theory or the proofs that accompany them, but, instead, with the translation from the language of mathematics (and logic) to everyday discourse. As a result, the early definition of computer viruses as elements which belong to a viral set had limited portablity to the world beyond the realm of logic, mathematics and computer theory. For the most part, all efforts to translate that language into a "plain text" definition failed. Adleman maintains, in fact, that such translations are impossible.[22]

What these difficulties in definition suggest is that the nature of virus programming is so closely tied to the basic operation of the computer that it is difficult, if not impossible, to distinguish virus code from other types of computer software. In short, there is something basic about virus programming that ties it to the very nature of the machines that these programs infect.

From Evolution to Infection

Five years, almost to the day, after Adleman and Cohen had defined computer viruses, in November of 1988, Robert Morris would unleash what would be called the "Internet Worm," a program which clogged the Internet so severely that it was considered crippled for a period of days. Morris's program was the first instance of a massive, network-wide infection.

The program, released on the ARPANET, spread throughout the then budding university, corporate, and government research networks. As Peter Denning describes it, the program "expropriated the resources of each invaded computer to generate replicas of itself on other computers, but did no apparent damage. Within hours, it had spread to several thousand computers attached to the worldwide Research Internet."[23] As the program spread throughout the network, it would tie up computing resources and eventually cause machines to grind to a halt, along the way the program would exploit known security holes and attempt to crack users' passwords. "Computers infested with the worm were soon laboring under a huge load of programs" and efforts to rid the system of the worm proved futile.[24] "Attempts to kill these programs were ineffective: new copies would appear from Internet connections as fast as old copies were deleted. Many systems had to be shut down and the security loopholes closed before they could be restarted on the network without reinfestation."[25] As to whether the rapid spread of the program was intentional, accidental, or the product of sloppy coding was the subject of much speculation. One thing was clear, however, Morris's worm program was unlike anything that the computer community had seen before. One of the primary issues that concerned computer security experts was that the program had been unleashed in a networked environment. Unlike previous viruses, which were predominately attacking PCs and Macintosh computers, this virus was built to travel through a network. PC viruses attacked stand-alone computers and were transferred by infected floppy disks, most likely, and through shared software. The Morris worm was different because it did not rely on a user running a program. The worm moved on its own, replicating quickly and spreading not only through the computer system, but also to any computer connected to that system on the network. The program was similar to what two Xerox PARC researchers, John Shoch and Jon Hupp, had described in a 1982 article–the first computer worm, a program which moved across the computer network under its own power, in Shoch and Hupp's example, performing useful functions.[26]

After Morris's Internet "attack," the computer community had considerable debate about what to call the program, a "worm" or a "virus"? After a series of debates, the term "worm" won out. The debate hinged on how one defined a program.

Looking at the code itself, the worm was significantly different from how viruses were programmed, making it a distinct type of program. Looking at the effects of the program, it looked much more similar to a computer virus than it did to any previously programmed worms. The analysis which centered on the program itself was more persuasive and as a result, the worm nomenclature stuck.

One effect of this definitional problem, as the Morris case illustrates, is the proliferation of discourses about computer programs. Rather than defining computer viruses as a distinct entity, programs which are not to be included in the category of viruses are given their own nomenclature and classification. Spafford, for example, lists a number of programs which are "self-reproducing or malicious," but are generally considered as distinct from viruses. These include "back doors," "trapdoors," "logic bombs," "worms," "Trojan horses," "bacteria," "rabbits," and "liveware."[27] Such definitional machinations mirror precisely the evolution of disease that Foucault identified in the eighteenth century. Viruses, by the very inability to define or classify them independently, produce an entire structure or "rational order" of diseases, which are systematically excluded from the logics and rationality of the discourse with which they co-exist. As a result, an entire discourse springs up out of the *inability* to define a computer virus. Computer viruses become defined by analogy. They are *like* other programs which are also different from what are accepted as "useful" or "normal" in computer programming. One only knows that computer viruses are unlike "normal" programs because they are *like* other dangerous programs which also differ from the norm. In Foucault's terms, the distance that separates one "dangerous" program from another "can be measured only by the *degree* of their *resemblance*," without reference to their histories or the logics which create and sustain them.[28] The computer virus is unable to be distinguished from other types of programming by definition. Instead, it is differentiated from programming by being *like* several other types of programs which are also to be excluded from the essential nature of programming, primarily because of the effects they produce.

In this moment of separation, what appears to be revealed is two "natures" of the computer. On one side is the "rational order" of technology, maintained by a narrative of progress, linearity and order. On the other side is that which is excluded from the rational order, that which is noise or blockage in that linear narrative progression. From that moment of difference, is abstracted a second order or rationality which defines what is to be excluded. Viruses and other "self-replicating" or "malicious" programs are endowed with their own sense of order or purpose, constructed in opposition to the rationality of the master narrative.

A related problem in defining viruses is their rapid rate of change and polymorphism. In order to stay ahead of anti-virus software such as scanners and inoculation programs, virus writers rely on constant change and development and the concept of polymorphism, the idea that a virus can take many different shapes or forms, not just one. The viruses are, themselves, "self-mutating," capable of changing their appearance and digital signatures to become undetectable. Change, then, both becomes a tool for virus writers, but also a concept that is encoded into the very structure of the virus itself.

This idea of change makes it nearly impossible to define viruses in such a way that they are clearly delineated from other types of useful programs. Instead, viruses are defined more commonly by the effect they have on the user's machine. More properly called "malware" (for *mal*icious soft*ware*), these programs are designed to cause harm to a user's computer. While these programs make up the minority of the computer viruses written, they have the highest visibility and inspire the greatest fear. As a result, they have come to represent the entire family of programs which more properly bear the name "virus."

Dissecting the Internet Worm: Virus Culture and Medical Perception

The notion of computer viruses are often tied to their biological counter-parts comes as no surprise. In a culture which often privileges simulation over reality, computer viruses are one of the few objects that have direct, profound, and immediate effects on the world. They are a reminder of the connection between the virtual and the real. But they also make plain, in often dramatic ways, the degree to which our culture has grown to rely on computers in nearly every context of human activity and interaction.

One of the primary responses to Morris's Internet worm came from researchers at MIT. The approach that MIT undertook to understanding the worm was to employ a kind of medical perception, literally, as Rochlis and Eichin (two of the MIT researchers who worked on the project) described it, "our group at MIT wound up decompiling the virus and discovering its inner details."[29] The connection to medical perception was made explicit both in their media releases and in the title of the paper they published in the *Communications of the ACM*, "With a Microscope and Tweezers." Donn Seeley, a University of Utah professor performed a similar analysis on one of the worm's key algorithms.[30]

The idea of a "post-mortem" analysis of the Internet worm firmly established the medical gaze as the primary epistemological frame of reference for the study of computer viruses. Just as Foucault had documented a shift in medical perception around the ability to open up corpses, the analysis of viral infection deployed precisely the same techniques and logics of "pathological anatomy." In Foucault's terminology, the institution of pathological anatomy was to become medicine's "most vital expression and its deepest reason."[31] The analysis of the Internet worm had three main approaches: an analysis of its structure, a chronicle of its spread, and a detailing of its algorithms.[32] Regardless of the decision as to whether or not to call Morris's program a worm or a virus, the means by which the program was apprehended and neutralized firmly entrenched medical perception and pathological anatomy as the means by which the program would be understood. The "host," a term which has meaning in both biological and computer science, is infected by a disease

Disease, since the eighteenth century, Foucault reminds us, became the subject of medical perception when it attached itself to the body. Corporeal objects provided a space for invention and refinement of the medical gaze. The body was, in Foucault's terms, "the concrete space of perception."[33] While the discourse that followed Morris's Internet worm incident invoked corporeal metaphors, the connection between the body and computer networks had already been developed a year earlier by Charles Cresson Wood. Wood argued that computer security necessitated a "new way of thinking" and "new reference models. Just as military advances were inspired by animals (birds inspired airplanes, bats inspired radar)," he argued, "the human body's immune system can inspire new advances in systems security."[34] The references that Wood suggests specifically are "vaccinations," "white blood cells," "antigens," "free radicals," "inflammation and fever" and "acquired immuno-deficiency syndrome (AIDS)."[35] Such suggestions document a whole-scale epistemological shift to a system of medical perception. In Wood's terms, "the human body can provide a wealth of security-relevant ideas."[36]

Wood's connection between the body and computer security was intended to compliment what he saw as a paradigm shift occurring in contemporary thinking about technology. That shift, positioned viruses within a biological discourse similar to Adleman and Cohen's understanding of networks, biology and evolution. Wood himself argues that "a systems security analogy to the human body's immune system is fully compatible with the now-underway paradigm shift from the information age to the age in which we holistically regard the planet earth and its subsystems as intricately linked webs of technologies and systems. . . the futuristic view of the world as one integrated organism."[37] Wood's analogy would introduce a model which would be developed in a number of different directions in the years that followed. In particular, Morris's worm program would re-configure the discourse of medical perception in a way which would at once deploy and transform Wood's model.

The transformation of this system of medical perception was made explicit in a series of connections put forward by ACM President Bryan Kocher in his President's Letter, published in the *Communications of the ACM* under the title "A Hygiene Lesson."[38] The letter was a response to the incidents surround Morris's Internet Worm. The discussion, rather than being about evolution, technology or ethics, was about what Kocher called an "electronic epidemic." Kocher calls for a particular kind of medical perception in the analysis of and protection against computer viruses. "I believe," he wrote, "that after many years of fruitless admonitions by the NSA, a way has finally been found to focus serious attention on system security, i.e. hygiene."[39] Kocher compares the spread of Morris's Internet Worm to the spread of cholera in Calcutta and, later, in Britain. The lesson, he argues, is that "just as in human society, hygiene is critical in preventing the spread of disease in computer systems. Preventing disease requires setting and maintaining high standards of sanitation throughout society, from simple personal precautions (like washing your hands or not letting anyone know your password), to large investments (like water and sewage treatment plants or reliably tested and certified secure systems."[40] Kocher's message received a mixed response, with large number of computer scientists maintaining that the issue was an ethical one, resisting the descriptions of "disease," "infection," and public health scares. Those, however, would be the terms which would capture both the media's attention and the public imagination. As Kocher concluded, "We must heed the public health warnings from NSA, practice personal systems hygiene, adhere to sanitary standards, and support the development of secure systems to keep the germs out."[41]

As some had predicted (notably Gene Spafford), there are dangers with "coopting terms from another discipline to describe phenomena within our own (computing). The original definitions may be much more complex that we originally imagine, and attempts to maintain and justify the analogies may require a considerable effort."[42] Spafford's concerns recognized only the problem in terms of the computer science community itself. The broader concern was one which would be realized as the discourse about viruses became an issue of public concern.

Medical Perception as Social Control

Cast as a public health threat, viruses sparked two discourses of social control. The first was a discourse which vacillated between arguments in favor of stiff legal penalties and those which advocated a more normative solution through the establishment of "a stronger and more effective ethical code among computer professionals" and "better internal policies."[43] In the late 1980s and early 1990s, however, these remedies were seen as inadequate for the simple reason that viruses were not judged to be "a sufficiently serious threat to the public welfare," even in the wake of the Morris Internet worm.[44]

The second discourse, which emerged from the Morris Internet worm debates which, instead, treated viruses not as legal or moral issues, but as biological and medical ones. The issue of hygiene became central to the discussion of computer viruses. The metaphor of hygiene (although many computer scientists would insists it is not metaphorical at all), is a powerful one. It articulates a threat and at the same time it taps into a well-established discourse of public health.

Historically, the discourse of hygiene has had a number of functions. Its primary function, however, was as a means of social control. Indeed, the very concept of the "police," Foucault argues, originated in part with enforcement of "general rules of hygiene" and in the eighteenth century, as the health of the population became increasingly important to the economic well-being, there emerged a "more general form of a 'medical police.'"[45] The role of the medical establishment, born out of responses to the epidemics of disease and plagues of the seventeenth and eighteenth centuries, developed into a "program of hygiene," which entailed "a certain number of authoritarian medical interventions and controls."[46] Specifically, Foucault argues, "the needs of hygiene demand an authoritarian medical intervention in what are regarded as the privileged breeding grounds of disease."[47]

This second discourse, which focused on disease, contamination and public health, transcended questions of legality or morality. Biological infections neither obey the law nor do they respond to or violate moral or normative codes. What they do provide, however, is a justification and warrant for extreme measures, particularly in relation to "sanitation." Those sanitary practices were located in two distinct domains. First, in the machines themselves, which Kocher had described in terms of hygiene. The solution, at least in part, rests in changing the conditions which allow the disease to grow unchecked. "The UNIX epidemic," he argued, "is like any other epidemic disease. It won't go away until the conditions that allow it to flourish are changed to prevent further infection."[48] Kocher's comments were met with a flurry of criticisms, mainly aimed at Kocher's failure to hold the virus's author responsible. As Narten and Spafford argued, for example, "computer viruses are created by persons deliberately circumventing known safeguards and actively seeking to spread infection."[49]

These two discourses, the legal/normative and the biological, appear to be in opposition, particularly around the question of agency and responsibility. They meet, however, in the discourse of hygiene. The discourse of hygiene allows for a new characterization of virus writers which fits both into the discussion of legal and moral responsibility and into the narrative of biological infection. Typical of such descriptions is Jan Hruska's characterization of virus writers (which she labels as "hackers") as "people analogous to drug addicts. They need their 'fix' and cannot leave the machine alone. Like addicts they seek novelty and new experiences. Writing a virus gives them this, but unlike addicts who get immediate relief after a fix, they are not usually present when the virus triggers and releases the payload."[50] The metaphors of drugs and additions (a primary public health threat in the early 1990s) are further developed by Hruska in his discussion of what he labels as "freaks." Freaks, Hruska writes, are "an irresponsible subgroup of hackers, in the same way that while some drug addicts remain reasonably responsible (and use sterile needles), others (psychopaths) become irresponsible (and share needles). Freaks have serious social adjustment problems and often bear general, unspecified grudges against society. They have no sense of responsibility or remorse about what they do, and are prepared to exploit others in order to achieve their aims."[51] In such a characterization, the discourse of hygiene is full of powerful references. In particular, the connection between drugs use and viral infection (particularly hepatitis and AIDS) from unclean or shared needles is further documentation of virus culture as a public health threat. It is a threat without specified agency. Although IV drug users are at great risk from infection, it is wrong to say that they *desire* infection. The public health threat is not found in their desire, agency or intention, but in their lack of hygiene.

The question of hygiene is transformed in Hruska's description into a lack of moral agency. The virus writer occupies the position of the psychopath, beyond moral or ethical conscience. The virus writer is unable or incapable of distinguishing between good and evil and between right and wrong and this lack of moral agency poses the more dire sort of threat to the public. As Hruska concludes, "The mentality of the freak virus writer is not unlike that of a person who leaves a poisoned jar of baby-food on a supermarket shelf. He delivers his potion, leaves and is untraced, and in his absence the victim falls."[52]

While these images are born of the discourse of hygiene, virus writers themselves see computers and the program that run on them from an entirely different vantage point. Rather than as a discourse of hygiene, virus writers have invoked a discourse of style as the means by which the challenge the meaning of technology, computers, computer culture and programming.

Virus Writers: Subculture and the Electronic Meaning of Style

Within the world of the computer underground there is a subculture devoted to the creation and dissemination of viruses. These programs, which range from code which prints humorous messages across the screen to programs which delete or destroy information, are produced and exist in relation to a clearly marked and defined subculture of virus writers. Unlike the image of virus writers as "high tech vandals," many of these programmers are often very talented and see virus writing as a social, cultural and political project. In contrast to the commonly held assumptions about virus writers (such as those described by Hruska above), I argue that the history and motives of this subculture are revealed in the conditions under which that subculture was born and evolved as well as in the shifting cultural and political contexts to which these programmers have responded. Accordingly, the creation of viruses is not merely a malicious act of vandalism or a senseless act of high-tech destruction, but, instead, functions as a means of subcultural signification and as a strategy for the preservation of a subcultural style in an age of increasing incorporation and commodification of underground computer culture. Like hackers, a distinct subculture which rarely overlaps with the world of virus programmers, virus writers create their own loosely affiliated groups, publish their own underground journals, and engage in competition and technological one-upmanship, constantly striving to outdo each other in feats of programming. As stated in the "The Constitution of Worldwide Virus Writers," published in *40Hex*, under the title MOTIVATION: "In most cases, the motivation for writing a virus should not be the pleasure of seeing someone else's system trashed, but to test one's programming abilities."[53]

In their programming, virus writers have adopted a sense of style, which I designate as "viral style" in an effort to mark themselves a being outside of the computer science community as well as resistant to the computer and anti-virus industry. Virus subculture, like most subcultures, is difficult to define, primarily because as Hebdige has argued, the meaning of subculture is itself always in dispute.[54] That dispute centers on the primary function of the subculture itself–the use of style to make and remake cultural meaning. Style is, for subcultures, "the area in which opposing definitions clash with most dramatic force."[55] What is at stake for virus writers is the meaning of not only the programs they create, but the broader social, cultural and political meanings of technology within contemporary society. In much the same way that viruses are related to the history and function of the computer, virus writers are inherently tied to the history of programming.

Virus programmers have a long history of sharing code and ideas, a process which is similar to the early computer programmers of the 1960s and 1970. What these hackers used to refer to as "bumming code," is a standard for development in the virus community. In the 1960s and 1970s, it was believed that software should be free, both in terms of its cost and in terms of its ability to be modified. If a programmer could find a way to make software run better, do more, or work more efficiently, then he or she had not only the right, but the responsibility to improve it. It became an ethic among programmers to share code, often times referred to as the "Homebrew ethic" (named for the Homebrew Computer Club, where it was widely practiced) or the "Hacker Ethic" (as documented in Steven Levy's 1983 book *Hackers*).[56] The idea behind the ethic was that each successive iteration of coding would improve and alter the program for the better, making it do more or run more efficiently. It would also, in the process change the code, arguably for the better. It is a model that virus writers adopted immediately, making their source code available and even distributing it freely in underground journals and through electronic bulletin board systems and the Internet. Indeed, most virus writers relate their experience of writing their first virus in a way which mirrors that ethic identically. As one virus programmer (Skism One) explains, after being infected with the Jerusalem virus himself, I "examined the copy of Jeru for months. Then one day I used a Hex editor to change the suMSDOs string to SKISM-1. Then I went to all the computers I could find and infected them. The next thing you know my friend shows me this list with my name on it."[57] The hex editor was able to change the name of the program, and attribute authorship to Skism One, but it didn't actually modify the code. In order to actually become a virus writer, he needed to learn how to create virus programs on his own. At that point, Skism One, following in the tradition of computer programmers before him, built his own virus from what he learned from Jerusalem, "Then - well I got into assembler and disassembly and I started to learn how to modify the code and all that. The next thing you know I had made my own virus from the scraps of Jeru."[58] Before long, he had learned enough to begin constructing his own code, "Then I guess I grew out of the scavenger mode and started writing my own shit, from scratch."[59] While the ethic is similar, the goals and desires of the computer programers of Homebrew and the virus writers of the 1980s and 1990s are markedly different.

As a subculture, virus writers present themselves as what Dick Hebdige identified as *noise*, "as an actual mechanism of semantic disorder: a kind of temporary blockage in the system of representation."[60] Subculture, Hebdige argues, manifests itself in *style* as an intentional form of communication whereby the cultural and social negotiation of signs takes place. Marked as a kind of deviant behavior, subcultural style is often times "incorporated" through the process of commodification which results in the "conversion of subcultural signs into mass-produced objects (i.e. the commodity form)."[61] The result is a "diffusion of the subculture's subversive power," (e.g the way in which "punk innovations fed back directly into high fashion and mainstream fashion" in the 1970s).[62]

The birth and growth of virus culture can be traced directly to the commodification of the computer in the form of the PC and as a response to the incorporation and dilution of computer culture which accompanied the mass marketing of the PC. In terms of technology, the mass appeal of the personal computer, particularly in the mid-1980s, produced a widespread incorporation of computer culture, taking the essentially subversive "hacker style" which demanded an intimate knowledge of these machines and how they worked and stripping it of its transgressive character for mass consumption. With the introduction and growth of GUIs (graphical user interface), computer users have become increasingly distanced from the machines and software that they use. As a result, the technology has been rendered increasingly opaque, even as it has become easier to use and more "user friendly."

Virus writers are reacting with a kind of digital violence to these transformations which have taken place in computer culture. The dynamics of the production of a commodified opaque technology have created two motives for virus production. First, viruses force the end user to become aware (or at least more aware) of his or her blind reliance or dependence on technology. In doing so, the threat of viral infection forces him or her to take note of the technology itself. The threat of viral infection forces the end user to understand how his or her computer works, to take precautions, to be aware of how viruses are spread and how to protect oneself. As one writer commented about the threat arising from the Microsoft Word Macro viruses: “Control is in your hands. Don’t panic. Take this as an opportunity to learn more about the features of the software you use, to test and verify any security features you plan to utilize and then to configure accordingly.”[63] To the virus writer, the philosophy is simple--there are risks associated with ignorance, especially with ignorance about technology. Typically, virus writers are more hyperbolic in their assessments. But these assessments also betray an underlying metaphor, consonant with earlier theories of viral infection, namely, evolution. As Dark Angel (one of the more vitriolic virus writers) sees it: “Virii [sic] are wondrous creations written for the sole purpose of spreading and destroying the systems of unsuspecting fools. This eliminates the systems of simpletons who can't tell that there is a problem when a 100 byte file suddenly blossoms into a 1,000 byte file. Duh. These low-lifes do not deserve to exist, so it is our sacred duty to wipe their hard drives off the face of the Earth. It is a simple matter of speeding along survival of the fittest.”[64] While many, even most, virus writers in the subculture do not harbor such malicious intent, few would disagree with the assessment of the typical computer user as an "unsuspecting fool" and many see viruses as a natural extension of computer programming and as a logical step in programming and the computer's evolution.

The second motive stems from the fact that virus writers see the broader cultural implications of such dependence on technology as well. They see the commodification of the computer and of computer culture as leading to the possibility of technological domination and viruses provide a sense of protection. Accordingly, the targets for viral infection are not just individual users. Instead, the inspiration for such coding comes from precisely the kind of blockage that is characteristic of subcultural style. As one virus writer explains, the typical virus writer "is usually just this angry kid who happens to be very clever, and what's going through his head when he codes this thing is how fuckin' cool it'll be when it starts blowing holes through the infrastructure of some industrial monolith like IBM, where a bunch of drones will start going bugfuck when everything stops working."[65] The idea of being able to "smash the system" has its roots as far back as John Brunner's sci-fi classic *The Shockwave Rider*, which tells the story of a computer programmer who liberates society from the tyranny of technology by releasing a virus-like program throughout the government's network. Like Brunner's hero, virus writers see themselves as maintaining the fragility of the system, keeping it in a state of precariousness such that its power over us is always limited by our ability to destroy it.

As the culture of virus writers evolved, journals like *40Hex* underwent transformations as well. *40Hex* started as a purely technical journal with the sole mission of disseminating virus code. By 1995, the journal had taken on an entirely different tone: "We are going to get a little bit more political then we used to be, but we will still keep cranking out the high quality technical information that you all enjoy. I would strongly recommend that you don't skip over the political parts of the magazine, because there are people who want to make laws that will affect every reader of this magazine."[66] The political message that *40Hex* was remarkably similar to that of the early 1960s and 1970s hackers. Predicated on freedom of information, the editors of *40Hex* would argue that efforts to legislate against virus production were misguided. They maintain it isn't the production of viral code, but its distribution which should be made illegal. The *40Hex* attitude towards viruses, also betrays a deeper philosophy about the place of viruses in networks. Viruses have become so deeply embedded in the network environment, they argue, that they no longer able to be contained. "Unfortunately, it is too late to start working on anti-virus writing legislation now. The damage has been done. The virus issue is fairly similar to the AIDS issue. You have to use protection, no matter what. There will never be an end to virii. Even if everyone stopped writing virii, the infection rate wouldn't decrease. I don't know of many people that get hit by the newer strains that have been coming out. Most people still get hit by Jerusalem, Stoned, and other 'classics'."[67]

The lessons of science fiction are a primary target for the contested meaning of technology between the computer science community and the computer underground. In the late 1980s, the emergence of "cyberpunk" signaled a shift in the ways that the underground was thinking about technology, and it would be reflection on Morris's Internet worm, which would lead to that realization. "More often then we realize, reality conspires to imitate art. In the case of the computer virus reality," Paul Saffo wrote, "the art is 'cyberpunk,' a strangely compelling genre of science fiction that has gained a cult following among hackers operating on both sides of the law."[68] What Saffo would find interesting was the "generation gap" between the computer science community (most of whom, including Robert Morris, had read and been fan's of Brunner's *The Shockwave Rider*) and the underground world of virus writers, who were more devoted to the work of writers like William Gibson. With respect to Gibson's cyberpunk classic *Neuromancer*, Saffo would report "I am particularly struck by the 'generation gap' in the computer community when it comes to *Neuromancer*: virtually every teenage hacker I spoke with has the book, but almost none of my friends over 30 have picked it up."[69]

Virus writers have learned an important lesson from the past. Watching the process of commodification, whereby computer style has been commodified and stripped of its potentially subversive force, virus writers have adopted a different sense of style which has made them more resistant to cultural incorporation. Accordingly, viral style represents a break from Hebdige's initial notion of subculture in an important respect. The culture surrounding viruses is a subculture which demands constant innovation and which accounts for mainstream culture's (and the industry's) ability to commodify and incorporate aspects of subversive style. Viral style is a response to a response, a *mutation* of style which is negotiating precisely the moment of incorporation of earlier computer style and culture by the mainstream.

With the mass-production of the personal computer and the widespread incorporation of computer culture, viral style emerged as a style that negotiates a *previously incorporated style*. Cognizant of the dangers and possibilities of incorporation itself, viral style is *self-replicating* and *polymorphic*, continually changing with each iteration. It is a style that enacts its own defense mechanism against incorporation. It is a style that is constantly evolving, reclaiming the earliest trope which defined the discourse of computer viruses themselves. The antecedents of virus culture, the impact of virus culture on computer culture generally, and the context which situates virus culture is both politically and culturally reactive to the incorporation of hacker style in mainstream, dominant culture.

Virus writers were not the only group to respond to the discourse of infection, public health and hygiene which followed in the wake of Morris's Internet worm. The discourse of industry and the mass media also seized on the opportunity to exploit the network-wide infection. In the wake of Morris's worm, public perceptions of viruses and virus writers were formed, and a multi-million dollar industry was launched.

The Idea of the Computer Virus in the Popular Imagination

The emergence of computer viruses in popular culture was coincident with an evolving discourse of AIDS and infection which had risen to the level of a public health epidemic. In this sense, the discourses of infect, contamination, and hygiene were narratives that had already been mobilized with dramatic effect and provided an immediate and highly charged context for understanding computer viruses.

Unlike other types of infection, AIDS was constructed in the popular imagination as a kind of smart virus. As Marita Sturken argues, the discourse of AIDS disrupted the conventional narratives of infection in part because AIDS was endowed with agency and intentionality, learning how to mutate in response to the body's immune system.[70] AIDS, Sturken argues has transformed the nature of viral infection in the popular imagination: "A virus is not 'alive,' according to science, yet neither is it dead; it can be killed. It is a 'bundle' of genes, an incohesive tangle. It 'contains instructions' but apparently did not write them itself. It is 'pure information,' yet information that acquires meaning only when in contact with cells."[71] Within this context, the transformation of the virus into the discourse of information, it becomes clear that just as computer science had coopted the discourse of biology, biological science was now turning to information sciences for its metaphors. The issues which most directly affected the discourse of how AIDS functioned had to do with how it learns, propagates, and mutates, properties which were well understood and easily modeled in self-replicating computer code and computer models and simulations.

While the body was the site of discourse for AIDS, the larger social body and networks are the space in which the discourse of computer viruses is localize. In the wake of the 1988 worm for example, Peter Denning would write: "Certainly the vivid imagery of worms and viruses has enabled many outsiders to appreciate the subtlety and danger of attacks on computers attached to open networks. It has increased public appreciation of the dependence of important segments of the economy, aerospace systems, and defense networks on computers and telecommunications. Networks of computers have joined other critical networks that underpin our society–water, gas, electricity, telephones, air traffic control, banking, to name a few."[72] It is a sentiment which would be picked up on by the press, media, and in the popular imagination.

Until the release of Morris's Internet worm, the concept of a computer virus was abstract, limited, and contained. Viruses were occasional programs which spawned a few variants each year. For example, prior to Morris's Internet worm, there were 11 known viruses for the IBM PC. In the year after the worm, that number nearly doubled (to 21) and in the next few years that followed, the number grew rapidly doubling, roughly, every ten months. Even so, the number of virus programs still numbered in the low hundreds until Microsoft released a version of the Word word-processing program which had a built in macro function. The macro function allowed users to embed commands inside of Microsoft Word documents, giving near complete control over file access, creation, and deletion to the Word program. In doing so, Microsoft created a new virus delivery system which made it possible to transmit viruses through text documents (rather than as executable code). It also allowed viruses to be created without any knowledge of assembly language or computer programming. Macros were their own high-level language, easily understood and quickly apprehended by even computer neophytes. It was a language that was designed to be simple to utilize.

The release of Microsoft Word would turn out to be the most significant event in virus production. By 1994, the year of Word's release, the number of viruses were in excess of 5,000, nearly all of the new additions took advantage of the Word macro function.[73] The trend has continued throughout the 1990s. In 1998, the number of known viruses in the wild was in excess of 18,000.[74] By April of 1999, roughly one year later, McAfee and Associates reported 40,000 known viruses.[75]

The very idea of a virus connotes illness, sickness, and even death. Interestingly, however, the majority of viruses don't cause significant damage. In a 1996 survey, viruses were found to be the "most common type of security breach," with half of the companies surveyed reporting "virus incidents," however only five percent of those incidents were described as having a "serious or significant impact."[76] What surveys routinely show is that industry spends huge sums of money protecting against virus attacks and attributes large losses to virus infection (usually first on the list of "sources of financial loss" for organizations reporting virus infections), but that a very small number of actual "attacks" account for most of the damage. Even recent reports of the top 10 viruses reported on the Internet, find that only one carries any "payload" at all.

Most viruses, instead, fall into the category of what could be considered "pranks," a long-time mainstay of both the mainstream and underground computer communities. Those pranks take on a heightened importance and danger in the context of a networked environment where it is difficult or even impossible to predict their effects. As Denning argued, "these software 'pranks' are very serious; they are spreading faster than they are being stopped, and even the least harmful of viruses could be life-threatening. For example, in the context of a hospital life-support system, a virus that 'simply' stops a computer and displays a message until a key is pressed, could be fatal."[77] The transformation of the computer virus from harmless prank, to dangerous contaminant, marked the moment at which is became possible to conceive of virus programming as criminal conduct. In doing so, the computer virus, became an object of legal and judicial scrutiny.

Notes

1. "Innerview: Insights from Pioneers of the Silicon Revolution", *Networker*, September/October 1995, vol 7, No. 1.

2. Michel Foucault, *The Birth of the Clinic: An Archaeology of Medical Perception*, (A.M. Sheridan Smith, Trans.) New York: Vintage, 1975, p. xi.

3. Leonard M. Adleman, "An Abstract Theory of Computer Viruses," *Lecture Notes in Computer Science* Vol. 403, *Advances in Computing–Crypto '88*, S. Goldwasser (Ed.), Springer-Verlag, 1990.

4. Deleuze and Guattari, *A Thousand Plateaus*. p. 10.

5. Deleuze and Guattari, p. 12.

6.Deleuze and Guattari, p. 12.

7.Deleuze and Guattari, p. 12.

8. Deleuze and Guattari, p. 7.

9. Gilles Deleuze, *Difference & Repetition*, New York: Columbia University Press, 1994, p. 70.

10. Sherry Turkle, *Life on the Screen: Identity in the Age of the Internet*, New York: Simon & Schuster, 1995, p. 41.

11. Scott Bukatman, *Terminal Identity: The Virtual Subject in Post-Modern Science Fiction*, Durham, NC: Duke University Press, 1993, pp. 69-100.

12. Eugene Spafford, "Computer Viruses," in *Internet Besieged: Countering Cyberspace Scofflaws*, (Dorothy Denning and Peter Denning, Eds.), New York: ACM Press, 1998, p. 74.

13. Spafford, p. 74. Benford's original idea for computer viruses and vaccines was published in May of 1970 in *Venture* magazine and later developed in Gerrold's 1972 novel *When Harlie Was One*. New York: Doubleday, 1972.

14. "Innerview: Insights from Pioneers of the Silicon Revolution", *Networker*, September/October 1995, vol 7, No. 1.

15. VIRUS-L FAQ

16. Spafford, p. 75.

17. For example see Adleman, "An Abstract Theory of Computer Viruses"; Fred Cohen, Ph.D. Thesis, *Computer Viruses*, University of Southern California, 1983; Fred Cohen, "Computer Viruses–Theory and Experiments" *Computers & Security* Vol. 6 (1), 1987, p. 22-35.

18. Fred Cohen, *Computer Viruses*, Ph.D. Thesis, University of Southern California, 1983, reprinted in part as "Computational Aspects of Computer Viruses," in *Rogue Programs: Viruses, Worms, and Trojan Horses*, Ed. Lance Hoffman, New York: Van Nostrand Reinhold, 1990, p. 331.

19. Cohen, "Computational Aspects," p. 331.

20. Cohen, "Computational Aspects," p. 331.

21. Adleman, p. 319.

22. Leonard Adleman, interview, April 2, 1999.

23. Peter Denning, *American Scientist*, March-April, 1989, p. 126.

24. Denning, p. 126.

25. Denning, p. 126.

26. John Shoch and Jon Hupp, "The 'Worm' Programs–Early Experiments with Distributed Computing," *Communications of the ACM*, 25 (3): 172-180, March 1982.

27. Spafford, pp. 75-78.

28. Foucault, *Birth of the Clinic*, p. 6.

29. Mark W. Eichin and John A. Rochlis, "With a Microscope and Tweezers: The Worm from MIT's Perspective," *Communications of the ACM*, vol. 32, no. 6, 1989, p.689.

30. Donn Seeley, "Password Cracking: A Game of Wits," *Communications of the ACM*, vol 32 no. 6, 1989, pp. 700-703.

31. Foucault, *Birth of the Clinic*, p. 124.

32. Spafford, Eichin and Rochlis, and Seeley, respectively.

33. Foucault, *Birth of the Clinic*, p. 9.

34. Charles Cresson Wood, "The Human Immune System as an Information Systems Security Reference Model," *Computers & Security*, vol. 6, 1987, pp. 512-13.

35. Wood, pp. 513-516.

36. Wood, p. 516.

37. Wood, p. 516.

38. Bryan Kocher, "A Hygiene Lesson," *Communications of the ACM*, 32(1), January 1989, p. 3.

39. Kocher, p. 3.

40. Kocher, pp. 3; 6.

41. Kocher, p. 6.

42. Eugene Spafford, "The Internet Worm Incident," in *Rogue Programs: Viruses, Worms, and Trojan Horses*, Ed. Lance J. Hoffman, New York: Van Nostrand Reinhold, 1990, p. 206.

43. Pamela Samuelson, "Can Hackers Be Sued for Damages Caused by Computer Viruses?" *Communications of the ACM*, 32 (6), p. 668.

44. Michael Gemignani, "Viruses and Criminal Law," *Communications of the ACM,* 32(6), p. 671.

45. Michel Foucault, "The Politics of Health in the Eighteenth Century," in *The Foucault Reader*, Paul Rabinow, Ed., New York: Pantheon, p. 278.

46. Foucault, "The Politics of Health," p. 282.

47. Foucault, "The Politics of Health," p. 283.

48. Kocher, p. 3.

49. Thomas Narten and Eugene Spafford, "ACM Forum," *Communications of the ACM*, Volume 32(6), June 1989, p. 674.

50. Jan Hruska, *Computer Viruses and Anti-Virus Warfare* (2nd Revised Edition), New York: Ellis Horwood, 1992, p. 64

51. Hruska, p. 64.

52. Hruska, p. 65.

53. "The Constitution of Worldwide Virus Writers," 40Hex, Number 5, Volume 2, Issue 1, File 005, February 12, 1992.

54. Dick Hebdige, *Subculture: The Meaning of Style*, New York: Routledge, 1979, p. 3.

55. Hebdige, p. 3.

56. Steven Levy, *Hackers: Heroes of the Computer Revolution*, New York: Dell, 1984.

57. "Interview with Skism One - A.K.A. Lord SSS (triple S)," *Hex40*, Vol. 1, Issue 2, 1992, file 004.

58. "Interview with Skism One."

59. "Interview with Skism One."

60. Hebdige, p. 90.

61. Hebdige, p. 94.

62. Hebdige, p. 95.

63. In Charles Platt, *Anarchy Online: Net Crime*, New York: Harper Collins, 1996, p. 149.

64. Dark Angel, "Dark Angel's Phunky Virus Writing Guide." Text file. n.d. (c. 1992).

65. Platt, p. 145.

66. GHeap, "Welcome to 40Hex Issue 12," *40Hex*, Number 12, Volume 3, Issue 3, File 000.

67. GHeap, "Response to a Letter from Paul Melka," *40Hex,* Number 8, Volume 2, Issue 4, File 010.

68. Paul Saffo, "Consensual Realities in Cyberspace," *Communications of the ACM*, Vol. 32(6), 1989, p. 664.

69. Saffo, p. 665.

70. Marita Sturken, *Tangled Memories: The Vietnam War, The AIDS Epidemic and the Politics of Remembering*, Berkeley: University of California Press, 1997, p. 245.

71. Sturken, p. 245.

72. Denning, p. 128.

73. VIRUS-L FAQ, section F1.

74.*Dr. Solomon's Anti-Virus Deluxe*, Edition 1.1, April 1998.

75. McAfee Virus Information Center, http://vil.mcafee.com/villib/alpha.asp, April 11, 1999.

76. The Information Security Breaches Survey 1996, DTI, ICL, UK ITSEC, NCC. Cited in Dorothy E. Denning, "Cyberspace Attacks and Countermeasures," in *Internet Besieged: Countering Cyberspace Scofflaws*, (Dorothy Denning and Peter Denning, Eds.), New York: ACM Press, 1998, p. 39.

77. VIRUS-L FAQ, http://webworlds.co.uk/dharley/anti-virus/vlfaq200.txt